GROWTH IN LENGTH

GROWTH IN LENGTH

EMBRYOLOGICAL ESSAYS

BY

RICHARD ASSHETON, M.A., Sc.D., F.R.S.

Trinity College, Cambridge

Lecturer in Biology at Guy's Hospital in the University of London
Lecturer in Embryology at the Imperial College
Lecturer in Animal Embryology in the
University of Cambridge

With 42 Illustrations

Cambridge :
at the University Press
1916

CAMBRIDGE
UNIVERSITY PRESS

University Printing House, Cambridge CB2 8BS, United Kingdom

Published in the United States of America by Cambridge University Press, New York

Cambridge University Press is part of the University of Cambridge.

It furthers the University's mission by disseminating knowledge in the pursuit of education, learning and research at the highest international levels of excellence.

www.cambridge.org
Information on this title: www.cambridge.org/9781107696679

© Cambridge University Press 1916

First published 1916
First paperback edition 2014

A catalogue record for this publication is available from the British Library

ISBN 978-1-107-69667-9 Paperback

Cambridge University Press has no responsibility for the persistence or accuracy of URLs for external or third-party internet websites referred to in this publication, and does not guarantee that any content on such websites is, or will remain, accurate or appropriate.

PREFACE

THE three Lectures forming the first part of this book were given as one of the "Advanced Courses in Zoology" in the University of London. They were delivered at Guy's Hospital on January 23rd, January 30th and February 6th, 1913, under the title "The Growth in Length of the Vertebrate Embryo." They are now published in accordance with the known intention of the author, and in response to the expressed wish of some of those who heard them.

Although my husband had destined this work for publication, he had not actually prepared it for the press; but the Lectures were typewritten and some corrections and notes had been added to them. For help in arranging them for print I am greatly indebted to Professor Stanley Gardiner, F.R.S., and to Professor J. P. Hill, F.R.S., who have most kindly read the MSS; and to the latter especially for valuable suggestions as to some alterations in the original wording, which while adapted for lecturing purposes might not always have been clear to the reader.

The figures have been drawn by Miss D. Thursby-Pelham from the sketches given in the MSS referring to lantern slides and diagrams displayed during the lectures.

The cost of publication is partly defrayed by a grant from the Royal Society.

The second part of the book consists of a reprint of a paper on the Mechanics of Gastrulation which appeared in the *Archiv für Entwicklungsmechanik der Organismen* in 1910. It is here included as the subject is kindred to that of the lectures and also because it is often in demand and further copies are unobtainable.

As these "Growth Problems" occupied my husband's studies and interests perhaps to a greater extent than did any other of the varied Biological work undertaken by him, this small memorial volume may be of special interest to those who knew him as researcher and teacher. It is now published in the hope that it may fulfil his constant wish to inspire others to delve into this, to him, fascinating field of research. It represents the summation of work and thought carried on for more than twenty years.

FRANCES A. E. ASSHETON.

GRANTCHESTER,
April, 1916.

CONTENTS

PUBLICATIONS BY THE SAME AUTHOR

"On the Development of the Optic Nerve of Vertebrates and the Choroidal Fissure of Embryonic Life."
Quarterly Journal Microscopical Science. Vol. xxxiv, 1892.

"A Re-investigation into the Early Stages of the Development of the Rabbit."
Quarterly Journal Microscopical Science. Vol. xxxvii, 1894.

"On the Phenomenon of the Fusion of the Epiblastic Layers in the Rabbit and in the Frog."
Quarterly Journal Microscopical Science. Vol. xxxvii, 1894.

"On the Causes which lead to the Attachment of the Mammalian Embryo to the wall of the Uterus."
Quarterly Journal Microscopical Science. Vol. xxxvii, 1894.

"The Primitive Streak of the Rabbit, the causes which may determine its Shape, and the part of the Embryo formed by its activity."
Quarterly Journal Microscopical Science. Vol. xxxvii, 1894.

"On the Growth in Length of the Frog Embryo."
Quarterly Journal Microscopical Science. Vol. xxxvii, 1894.

"Notes on the Ciliation of the Ectoderm of the Amphibian Embryo."
Quarterly Journal Microscopical Science. Vol. xxxviii, 1896.

"An Experimental Examination into the growth of the Blastoderm of the Chick."
Proceedings of the Royal Society. Vol. lx, 1896.

"An account of a Blastodermic Vesicle of the Sheep of the seventh day with twin germinal areas."
            Journal of Anatomy and Physiology, April, 1898.

"The Development of the Pig during the first ten days."
            Quarterly Journal Microscopical Science. Vol. XLI, 1898.

"The Segmentation of the Ovum of the Sheep, with observations on the Hypothesis of a Hypoblastic Origin for the Trophoblast."
            Quarterly Journal Microscopical Science. Vol. XLI, 1898.

"On Growth Centres in Vertebrate Embryos."
            Anatomischer Anzeiger. Vol. XXVII, 1905.

"The Morphology of the Ungulate Placenta, particularly of that organ in the Sheep, and notes upon the Placenta of the Elephant and Hyrax."
            Proceedings of the Royal Society. Vol. LXXVI, 1905.

"On the Foetus and Placenta of the Spiny mouse, Acomys cahirinus."
            Proceedings of the Royal Society. Vol. LXXVI, 1905.

"Certain Features characteristic of Teleostean Development."
            Guy's Hospital Reports. Vol. LXI, 1907.

"The Development of Gymnarchus niloticus."
            Budgett Memorial Volume, University Press, Cambridge.

"Report on sundry Teleostean Eggs and Larvae from the Gambia River."
            Budgett Memorial Volume, University Press, Cambridge.

"The Blastocyst of Capra, with some Remarks upon the Homologies of the Germinal Layers of Mammals."
            Guy's Hospital Reports. Vol. LXII, 1908.

"A new species of Dolichoglossus."
            Zoologischer Anzeiger. Vol. XXXIII, 1908.

"Professor Hubrecht's paper on the Early Ontogenetic Phenomena in Mammals."
Quarterly Journal Microscopical Science. Vol. LIV, 1909.

"Tropidonotus and the Archenteric Knot of Ornithorhynchus."
Quarterly Journal Microscopical Science. Vol. LIV, 1909.

"The geometrical relation of the Nuclei in an Invaginating Gastrula (e.g. Amphioxus) considered in connection with Cell Rhythm and Driesch's conception of Entelechy."
Archiv für Ent. Mech. der Organismen. Vol. XXIX, 1910.

"Variation and Mendel."
Guy's Hospital Reports. Vol. LXIV, 1910.

"Loxosoma loxalina and Loxosoma saltans, a new species."
Quarterly Journal Microscopical Science. Vol. LVIII, 1912.

"Gastrulation in Birds."
Quarterly Journal Microscopical Science. Vol. LVIII, 1912.

"A review of Dr Beard's book on the Enzyme Treatment of Cancer and its scientific basis."
Guy's Hospital Gazette, February, 1912.

"Fission of the Embryonal Area in Mammals."
IXth International Zoological Congress. Monaco. 1913.

(With Dr A. Robinson.)
"Formation and Fate of the Primitive Streak."
Quarterly Journal Microscopical Science. Vol. XXXII, 1891.

(With Dr T. G. Stevens.)
"The Elephant's Placenta."
Quarterly Journal Microscopical Science. Vol. XLIX, 1905.

THE GROWTH IN LENGTH OF THE VERTEBRATE EMBRYO

LECTURE I

The mode of growth of the main axis of the Vertebrate Embryo has been for a good many years a subject of controversy among Comparative Anatomists and Embryologists. The problem is not such an easy one as it seems at first sight to be.

From a zoological point of view it is a very important question; because according to the way in which we are disposed to answer it we shall form our ideas as to the relation which the main axis of the vertebrate body bears to the main axis of the bodies of animals of other phyla. Also upon this problem depends to a large extent the view we take about metamerism. An Annelid is a distinctly metamerically segmented animal—so is a Vertebrate. Is this similarity a case of homology or of convergence? And what is the relation that the metamerism of the one bears to that of the other?

The problem of the growth in length of the Vertebrate Embryo may be said to have originated with the essay by Prof. Wilhelm His published in the year 1874 on the formation of the trout embryo ("Ueber die Bildung des Lachs Embryo," *Sitzungsberichte der naturforschenden Gesellschaft zu Leipzig,* 1. Jahrgang (1874), S. 30); the theory of the mode of growth in length of the embryo enunciated in this paper was expanded in subsequent papers, among which may be mentioned his well-known work *Unsere Körperform und das physiologische Problem ihrer Entstehung,* published in the same year.

The theory which His then put forward has been termed the *concrescence theory.* It is with respect to embryos of Teleosteans and Elasmobranchs that one is most inclined at first sight to adopt this theory of the growth in length of the Vertebrate.

A. V. E.   1

Oddly enough it is also to embryos of this type that those who oppose this view can best turn for their experimental evidence of a destructive character against the theory.

The theory of concrescence is as we shall see more difficult to combat with reference to the Amniota, for in them the theory confronts us in a more subtle form.

The eggs of most of the members of these two classes—the Elasmobranchs and Teleosteans—are large and are heavily laden with food yolk.

After the egg of an Elasmobranch has undergone a partial or meroblastic segmentation, a little disc or cap of cells is formed upon one pole. This we may call the upper pole, because in most cases, though not in all, the egg floats with the pole bearing this cap uppermost.

The cap of cells soon begins to expand and to become differentiated into distinct layers, an outer or upper one which is a continuous membrane, and is called the ectoderm or epiblast; and an inner tissue composed of loose cells and a yolk mass which together may be called the endoderm or hypoblast. This inner layer is continuous with the outer one peripherally, but at the posterior end it is continued inwards for a short distance and then becomes continuous with the yolk mass. The whole disc derived from the segmented region of the egg is called the blastoderm or blastodisc, the rest of the egg is called the yolk mass. This is the gastrula stage, and the narrow shallow cleft, which in the Elasmobranch extends under the edge of the blastoderm, is really the primitive gut cavity, and it may be called the archenteron. The edge of the blastoderm, where the ectoderm and endoderm are continuous with each other is therefore the blastopore lip; and the great yolk mass must be regarded as a piece of the floor of the gut projecting out of the gastrula mouth.

The diagram, Fig. 1, is really a compromise between the Elasmobranch and the Teleostean condition. In the Elasmobranch there is a much greater difference in size between the blastoderm and the yolk mass than in the Teleostean.

The Teleostean furthermore is complicated by the presence of an outer envelope which is split off from the ectoderm layer of cells. This layer is at first continuous with the yolk, but at a later stage it is also split off from the superficial layer of yolk. This outer investment therefore prevents the existence of an actual passage from the cleft-like gastrula cavity to the exterior, so that the archenteron is not very obvious in this class of fishes at this stage.

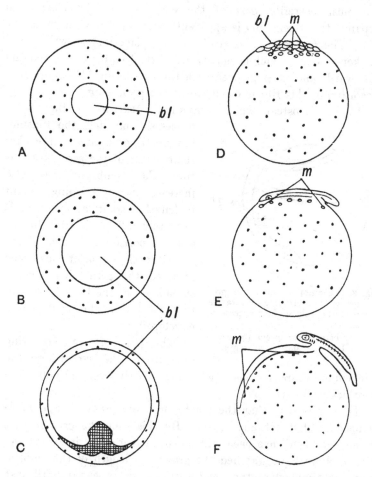

Fig. 1. Diagram to show the growth of the blastoderm over the yolk.

A, B, C, the egg as seen from above; *D, E, F*, the same stages seen from the side; *bl* blastoderm; *m* merocytes; the yolk is represented by dots.

As development proceeds, the blastoderm in each case becomes expanded and it gradually surrounds the whole yolk; but long before it does so there appears a thickening of the outer layer just inside the blastopore rim, and the rim itself also is thickened about this region. The inner edge of this thickened knob is the anterior end of the future neural tube, and the rim is therefore the most posterior part of the embryo. As the blastoderm expands this thickening is apparently extended inwards from the rim. The process that is really taking place is a growth of the thickened area—which is usually called the "embryo"—extending backwards and keeping pace with the expanding blastopore rim. Accordingly this rim is often called the germ ring.

In the Teleostean, which gives a much truer idea of the growth processes than does the Elasmobranch, the germ ring soon reaches the equator of the egg. By this time the "embryo" has also increased a corresponding amount in length, and its anterior end is definitely differentiated into the anterior portion of the fish, Fig. 2.

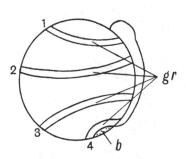

Fig. 2. Diagram to show the progress of the germ ring as it travels over the yolk at four successive stages.

b blastopore; *gr* germ ring; 1, 2, 3, 4, successive stages.

After the equator has been attained the germ ring quickly spreads on to the lower pole and in a short time the whole egg is enveloped.

The part of the germ ring opposite to the part where the thickening occurs, travels much more quickly over the yolk than the thickened part.

Thus it is seen that the blastopore becomes more and more reduced until it actually closes. By this time the greater part of the "embryo" has been laid down, in fact all of it except the tail; and we have watched the growth in length of the embryo from the time when the head region first appeared until just before the tail begins to grow. The question is how has this growth taken place?

When once an embryo is formed it may increase either in

general size, or in length only; it may increase as a whole, or in part; in any case the increase comes from a general multiplication of its cells, that is to say, it is the result of a general interstitial growth.

Or again there may be certain definite centres which continue for some time as centres of indifferent cell proliferation, and which give rise to growth in length of parts, as for instance, in the case of the limbs, or of the body as a whole.

Now in the particular case which we are considering, namely the growth in length of the fish embryo, it might be possible theoretically to ascribe the change from the condition seen in the earlier stage *A*, Fig. 3, to that of the later stage *B*, Fig. 3, as being due to one of three causes.

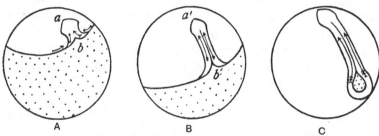

Fig. 3. Concrescence theory, after His.
The arrows indicate the direction in which the edges are approaching each other; the dotted area represents yolk.

1. It might be due to a general interstitial growth whereby *a*—*b* becomes *a'*—*b'*. In this case the foundation of the whole embryo would be considered as having been definitely laid down between these two points from the beginning.

This explanation however is probably not correct, because, as is perfectly well known, in all vertebrate embryos the anterior part of the embryo develops long before the hindermost. In Fig. 3 the earlier stage shows only the rudiment of the fore part of the brain. The rest of the brain and the spinal cord and the protovertebrae are added on subsequently, in regular succession.

Experimental evidence is also against this hypothesis.

2. Or, we may regard the advancing rim as being an area of special cell proliferation, and as giving rise to the more posterior part of the embryo as it passes backwards over the yolk.

We shall find that in the Anamnia the experimental evidence supports this view.

3.   Then again it is possible theoretically to hold the view that the main axis of the embryo is really formed by the coalescence of the lips of the rim—each half side of the rim representing one-half of the embryo.   This view is known as the concrescence theory.

Since the rim is regarded as the lip of the blastopore (the uncovered yolk as a plug filling up the blastopore) then, according to the concrescence theory of vertebrate development, the blastopore is supposed to fuse by the concrescence of its lateral lips, each lip giving rise to the corresponding half of the embryo, Fig. 3 *C*, and the blastopore finally closing.

According to this theory the relation of the main axis of a Vertebrate to the blastopore is very different from that which it has if we adopt alternative No. 2.   According to the concrescence theory the relation of the main axis to the old blastopore of coelenterate days—the gastræa stage—is the same in Vertebrates as it is in Annelida and Arthropoda.   If we adopt No. 2, the main axis is in a direction at right angles to what it is in Annelida and Arthropoda.

This theory of concrescence started by His in 1874 has been brought forward time after time by embryologists in newer and more subtle forms.   It has received support from many authorities, and has been vigorously opposed by others.

GROWTH CENTRES.

Quite apart from the question of concrescence—or the heresy of concrescence, as I would prefer to call it—there is another question of considerable importance, and of especial interest with respect to the difficult question of gastrulation and the formation of the germ layers, the ectoderm and endoderm.

If we look at an early stage of the trout embryo we see a perfectly circular disc of cells.   Later the rim thickens along the border which will ultimately be the posterior and dorsal part of the embryo and rim.   But what about the thickened area just within this rim? Fig. 1, *C*. Is that derived from the rim, or is it formed so to speak *in situ* on the blastoderm?   And what about the fate of the blastoderm just in front of this knob? Does not that give rise to part of the embryo? There can be no doubt about the answer.

All parts of the blastoderm of a Teleostean or an Elasmobranch become parts of the embryo.

Thus we have a part of the embryo formed from the layer of cells which exists as a result of segmentation, and another part which comes into being by the activity of the germ rim whether by concrescence or not.

That is to say we have two definite and in a sense independent centres of growth, one of these comes into being as the result of the fertilisation of the egg,—and the other comes into being subsequently.

These may be termed the *primary growth centre* and the *secondary growth centre*—or to use terms which I suggested some years ago, the *protogenetic* and *deuterogenetic* growth centres. The activities of these centres may be called protogenesis and deuterogenesis respectively.

In order to understand the action of these centres one must have a clear idea of the formation of the germ layers and of the process and meaning of gastrulation.

I propose therefore to describe in some detail the formation of the germ layers in the Amphibia.

Rana temporaria.

An egg of *Rana temporaria* is pigmented to a greater or less extent. There is much variation, in some the lower hemisphere is almost quite devoid of pigment, in others it is pigmented but is slightly lighter near the lower pole.

The whiter pole is the more heavily laden with yolk and it floats downwards in water. To all appearances the egg is radially symmetrical. Soon after fertilisation a curious crescentic whitening of part of the pigmented area can be noticed, this is called the "grey crescent."

It is said

(1) That this is always immediately opposite to the point at which the spermatozoon has entered;

(2) That the future sagittal plane of the animal coincides

approximately with the centre of this crescent and that of the egg.

Segmentation as a rule follows the radial type.

A pit begins on the upper pole of the black hemisphere, and passes round as a furrow dividing the egg into two equal segments, Fig. 4 *A, B* and *E, F*. Very soon, after about two hours or less according to the temperature, two furrows start from the centre of this furrow in the black hemisphere and gradually extend so as to divide each segment again into two, so that we have four segments of equal size all lying in the same plane, as in Amphioxus.

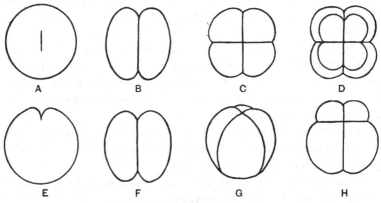

Fig. 4. Egg of *Rana temporaria*.

A, B, C, D, diagrammatic views of segmentation as seen from above; *E, F, G, H*, corresponding stages seen from the side.

If we compare several views of the lower pole we shall find that there are several varieties of mode of meeting of the furrows at the lower pole. Also we shall notice that the spot where the furrows meet is not in the centre of the white hemisphere.

The next set of furrows is horizontal. These may be seen commencing from the edges of the two previous furrows and eventually meeting one another, so cutting the four segments into eight.

In comparing this stage with Amphioxus we note that it corresponds to the third division plane in the radial type of Amphioxus, but is much nearer to the upper pole. There is thus a greater difference between the micromeres and macromeres.

The radial character is continued in some cases by the division in vertical planes, of each of the four micromeres, and at a slower rate, of the four macromeres, Fig. 5 *A*. But sometimes a distinctly bilateral phase is entered upon, Fig. 5 *B*.

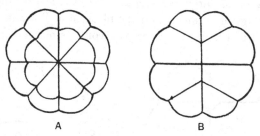

A   B

Fig. 5   *A*, Radial division; *B*, Bilateral division.

After this stage the segmentation proceeds less regularly and the bilateral symmetry if established is soon lost. As early as the eight segment stage a space is apparent between the segments very much nearer the upper pole than the lower. This is the first sign of the blastocoel or segmentation cavity. At about the sixteen segment stage this cavity is quite enclosed, and we have a condition comparable to that of Amphioxus, the chief difference being that the segments are more uneven in size, fewer in number and correspondingly larger, so that the blastocoel is correspondingly smaller, Fig. 6 *B*. Segments are now formed by tangential planes of division, so that the wall becomes first double and then many-layered, and at the close of segmentation we have a blastula very characteristic of Amphibians in which there is an upper pole of very small black cells and a lower pole of considerably larger and whiter cells, bounding between them the eccentrically placed segmentation cavity, Fig. 6 *D*.

There is a stage during the earlier phases to which I would draw special attention. In Fig. 6 *C* it will be noticed that there are very few division furrows extending to the lower pole, and no horizontal ones. This is of course due to the fact that the horizontal planes always result in the production of a small and a large segment, owing to the greater amount of yolk in the lower part.

In Rana we have a good example of an egg which is holoblastic.

but in which gastrulation can be said to be produced by invagination in a very small degree only.

Gastrulation is produced by a splitting among the segments, caused by a differentiation of an upper layer of small segments above a layer of larger segments. We have seen that in the advanced stage of segmentation the blastula of Rana has a many-layered wall of which the upper part consists of smaller cells and

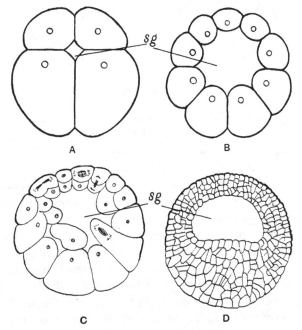

Fig. 6.   Frog's egg.   Segmentation.

A, 8 cell stage; *B*, 16 cell stage; *C*, 32 cell stage; *D*, stage with segmentation
cavity (*sg*) fully developed.

the lower of larger cells, Fig. 6 *D*, with a zone of more or less abrupt passage between the two; this will be seen in sections to vary slightly according to the radius along which the sections are cut.

At this stage, on the surface, one sees that the upper pole is formed by small black cells and the lower pole of whiter and larger cells. There is no sharp line between the two, and yet there is sufficient difference to enable one to speak of an advance of the small black cells into the region previously occupied by the large white cells. This advance is an advance of a zone of differentiation; it is certainly not an actual movement of black cells over the white, in Rana, though it may be so in some other Amphibia. Thus there is a tendency during segmentation for the black area to increase at the expense of the white, for pigment is deposited as the larger cells segment into smaller ones. *Gastrulation* is first rendered evident by the appearance of a slight groove running parallel to the equator of the egg but some distance below it, Fig. 7.

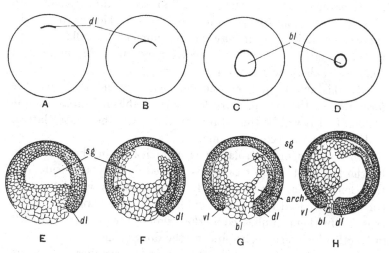

Fig. 7. Diagram to illustrate the external views and corresponding internal changes during gastrulation in the frog.

A, B, C, D, external views; *E, F, G, H,* median sections; *arch* archenteron; *bl* blastopore; *dl* dorsal lip of blastopore; *vl* ventral lip of blastopore; *sg* segmentation cavity; in *H* the segmentation cavity has almost disappeared.

We will now very briefly look at the general characters as one sees them in surface view and sections, without at first considering the mechanism.

If we examine the groove in section it will be noticed that there are signs of pressure, the cells are compressed and there is more pigment deposited between the segments at this point than elsewhere.

If we watch the surface we find, Fig. 7 *A—D,* that the groove which at first appeared almost straight and transverse, soon lengthens and becomes crescentic, and finally its horns meet and we have a white patch of larger segments surrounded by a dark sharply defined rim. The white patch is formed by a plug of yolk cells which almost completely blocks the blastopore, but all round the plug the archenteric cavity is visible as a chink between the plug and the lip rim. This rim is the blastopore lip.

Examination of vertical sections, Fig. 7 *E—H*, shows that a narrow slit, continues inward from the groove parallel with the surface and some half-a-dozen cells below it. The slit gradually lengthens inwards from the groove and eventually expands into a cavity as the blastopore forms and closes. This cavity is the future gut cavity, and it may for the moment be called archenteron, though as we shall see it corresponds to rather more than what we shall eventually term archenteron.

By this act of splitting the gut cavity is initiated, and the formation of the two primary layers is established, firstly in the region of what will be the future dorsal surface, then in the lateral and lastly in the ventral region.

The process of gastrulation is complicated in the Amphibia by the fact that the gut cavity is formed dorsally a considerable time before it is formed ventrally—or perhaps it would be more correct to say that it is formed anteriorly sooner than posteriorly. It is still further complicated by the origin of the process of growth in length which also begins first on the dorsal side.

Experimental observations show quite unmistakeably that as soon as the blastopore lip is formed it begins to grow over the yolk, and does not cease to grow in the dorsal region until the full complement of segments have been formed. The blastopore lip constitutes a ring-like area of proliferating tissue which is continuously adding on new material to all previously existing material with which it is in contact. It forms the deuterogenetic centre.

The series of sagittal sections, shown in Fig. 7, indicate the character of the changes which take place internally and are accompanied by surface views.

The slit seen in *F* running in from the groove shown in surface view in *B* widens out in *G* and *H* to form a capacious chamber which is the archenteron—the first formed part of the gut cavity.

The segmentation cavity which is so large and conspicuous an object in *E* becomes smaller and appears to be obliterated by gradual compression owing to the advance of cells caused by the expansion of the gastrulation slit.

At the same time it must be noted that there is some uncertainty as to whether the whole of the segmentation cavity is obliterated in this way, or whether some of it may not remain and become merged with the new archenteric cavity derived from the slit, and thus in fact help to form the future gut cavity.

There is no doubt that the wall between the original segmentation cavity and the slit cavity becomes very thin. According to Hertwig this wall breaks down in the frog. According to Brachet it breaks down sometimes in *Rana temporaria* though not always, but in Siredon it always does break down, as according to Brauer it does in Hypogeophis. Hence we may say that in Amphibia the segmentation cavity either disappears by compression of its walls or it may become confluent with the anterior region of the archenteron.

To compare this condition with Amphioxus we may refer to the diagrams of sections drawn in corresponding positions. It will be seen that a certain amount of invagination occurs. A mark placed in the yolk plug at X should appear within the gut cavity at a later period, Fig. 8.

There are no essential differences with respect to gastrulation in the many genera of Anura and Urodela which have now been studied.

In the case of the Gymnophiona however we have an egg which is distinctly nearer to the meroblastic type than is that of either of the other two Amphibian orders. It is almost a centro-lecithal egg.

In the egg of Hypogeophis, which has been very well studied and described by Brauer, the segmentation is partial; but there is a very rapid superficial differentiation so that a circular blasto-pore is formed very much as in the other Amphibia. The segmentation at one time shows a cap of small cells upon a mass of still unsegmented yolk, with here and there a nucleus near to the segmented pole. There is no very well marked segmentation cavity, but small cavities appear here and there among the small cells, Fig. 26 *A*. Brauer seems to draw a sharp distinction between "animal cells" and "vegetative cells." I cannot see the importance of so doing. All his figures show that the superficial layer of the whole egg very quickly becomes segmented or at any rate it becomes covered with a layer of cells. Possibly this may be to a certain extent due to a sliding of the animal cells over the surface, but it is far more probably due to a progressive differentiation, as it is in the frog.

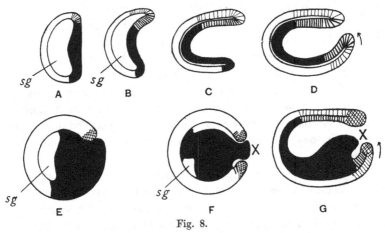

Fig. 8.

A, B, C, D, gastrulation stages in Amphioxus. After MacBride, 1910. *E, F, G*, corresponding stages in *Rana temporaria*; the white parts represent protogenetic ectoderm; the black protogenetic endoderm, and the thick and thin lines deuterogenetic endoderm and ectoderm respectively; *sg* segmentation cavity.

However that may be, a stage is eventually reached when there is a circular blastopore established, with lips continuous all round and with a superficial layer of cells on the outside and an inner cell mass, Fig. 27, p. 49.

In this stage the cavities which represent the segmentation cavity, become confluent with a space caused by splitting between the cells that radiate inwards from the blastopore lip. Thus in Hypogeophis we have at this stage the future gut cavity well defined into two regions: that which was in existence actually before or at the time of the beginning of the blastoporic ingrowth, and that which has been produced in connection with the growth of that edge.

Thus in Rana, Siredon, and Hypogeophis one sees very plainly from anatomical observations that the gut cavity is due to two processes.

A point to be noted is that when as in Amphioxus and Rana there is a real segmentation cavity, namely a cavity produced by the geometrical arrangement of the early segments, then this cavity, as in Amphioxus, is roofed over by cells which will eventually become ectoderm cells, and its floor is formed by cells which will eventually become endoderm cells.

Before it becomes converted into gut cavity there is a movement of segments so as to line the ectodermic roof. This movement in part is no doubt due to pressure caused by the expansion of the gut cavity, though partly perhaps to the pressure produced by the growth backwards of the blastopore lips. It can be taken as representing a modified form of invagination.

Invagination such as one sees in a yolkless egg (Amphioxus) is impossible—therefore gastrulation is affected by some other means.

It is doubtful to what extent the gut cavity is due to the splitting referred to above.

This is rather an important matter; because all the part of the gut cavity which is produced *either by splitting or by any modification of invagination,* or *by conversion of the segmentation cavity* (*so called*) clearly differs from that which is produced by the GROWTH BACK OF THE BLASTOPORE LIPS. *The former is in all cases protogenetic,* THE LATTER IS DEUTEROGENETIC.

The distinction between the two parts of the gut cavity is quite distinctly seen in Axolotl and Hypogeophis, as Brachet's figures of the former and Brauer's of the latter will demonstrate.

In Rana it is a little more uncertain. I have often tried to test this point by experiment in *Rana temporaria* and I think it can be done but it is not easy. It might be done by inserting fine hairs into the yolk cells just behind the dorsal lip of the blastopore and then tracing them after growth in length has continued for some time.

It seems quite possible that the greater part of the gut cavity of protogenetic origin is derived from the conversion of the segmentation cavity into gastrula cavity by the movement of those cells to which we alluded in the preceding paragraph. But in any case there is undoubtedly some splitting, even if it is only such as is sufficient to make the cavity derived from the segmentation cavity continuous with that derived from the overgrowth of the blastopore lip.

It might be well to restrict the term *archenteron* to the protogenetic part of the gut cavity and to call the deuterogenetic part *metenteron* ("neo-enteron" of de Lange), Fig. 9.

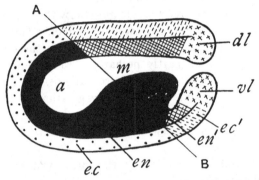

Fig. 9.   Rana.   Diagram to show protogenetic and deuterogenetic tissues with a dividing line *A*, *B*.

ec, en protogenetic ectoderm and endoderm: *ec′, en′* deuterogenetic ectoderm and endoderm; *dl* dorsal lip of blastopore; *vl* ventral lip of blastopore; *a* archenteron; *m* metenteron.

At first sight a difficulty arises because the cavity in question seems to be roofed in by deuterogenetic, while it is paved by protogenetic tissue. Really however the portion of the floor which corresponds to the deuterogenetic roof is the little piece near the letter *v*, Fig. 10.   The yolk plug is the hypertrophied floor of the protogenetic area or true archenteron, but it is projected out temporarily through the blastopore mouth.   At a later stage the yolk plug is actually withdrawn—this in itself may perhaps be an invagination process.

The reason why the roof of the metenteron is so much longer than the floor, is that the dorsal lip by which the roof is formed comes into being so much sooner in the frog than does the ventral lip.

For a time the whole annular area of deuterogenetic activity continues, with the result that the metenteron is increased in length and new tissue is added on to the back, sides, and ventral walls of the animal, and to all organs which at this time are present, and thus the whole embryo, gut, nervous system, notochord, mesoderm, increases in length—by the addition of new material posterior to all previously existing tissues.

The Tail.

After a certain time however and again as a result of the activity of the deuterogenetic centre, the lips of the blastopore approach and coalesce, and this is followed almost at once by the dying out of the ventral and ventro-lateral parts of that area of cell production.

The dorsal region and dorso-lateral region however continue their growth for a considerable time longer, but the chief parts with which they are now continuous are the neural tube, the notochord, and the dorsal and dorso-lateral parts of the ectoderm and mesoderm.

The metenteron is on the whole more ventral, so that there is no great tendency for the gut cavity to be continued backwards. Also none of the ventral or ventro-lateral part of the body wall is continued backwards. Growth in length ceases with regard to these parts.

In some Amphibians (some Urodela) the blastopore partly closes by the coalescence of its lateral lips, leaving the extreme ventral and the extreme dorsal parts open. The ventral part remains open as the anus after the dying out of the ventral region of the deuterogenetic centre. The dorsal part becomes involved in the folding up of the neural folds which adjoin its anterior and antero-lateral margins and it forms in that way a temporary communication between metenteron and neural tube. This is called the *Neurenteric canal.*

In Rana and other Anura the ventral part of the blastopore closes along with the middle part; but the anus forms in the closed area and, as the deuterogenetic area dies out almost immediately, no growth in length occurs here.

Thus the dorsal part alone continues to grow and to add on more and more segments, so producing the chordate tail, which released from the encumbrance and discomforts of a digestive system has led to the wonderful success achieved by the vertebrate phylum.

It is necessary to realize that the metenteron has a wide roof and narrow floor, because otherwise it is difficult to interpret sections taken vertically and transversely through the egg, e.g. along the line *A—B*, Figs. 9 and 13.

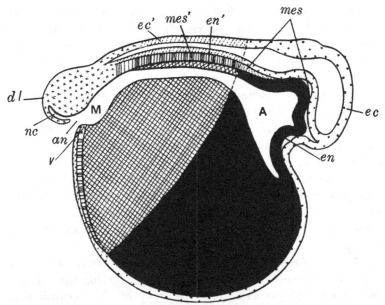

Fig. 10 after de Lange, 1912.   Diagram of embryo of Megalobatrachus.
an anus; *A* archenteron; *M* metenteron; *ec* protogenetic ectoderm; *en* protogenetic endoderm; *dl* dorsal lip; *v* ventral lip of blastopore; *mes* protogenetic mesoderm; *mes′* deuterogenetic mesoderm; *nc* neurenteric canal; *ec′* deuterogenetic ectoderm; *en′* deuterogenetic endoderm.

The deuterogenetic tissues have in the Frog a more compact and less yolky character than the protogenetic tissues.   The two regions of course merge one with the other along the lateral walls.

Thus we get the appearance of a special dorsal plate forming the nerve plate and notochord, Fig. 10.   It was this appearance which caused Lwoff in 1894 to describe what he termed the ectoblasto-genetic plate as a turning in of the ectoderm to form the notochord and the roof of the gut cavity.

De Lange has found evidence of these two growth centres in *Megalobatrachus japonicus* which quite agrees with what we find in Rana and Hypogeophis.

De Lange suggests a further term tritogenesis to describe the outer growth of the tail. But clearly the growth of the tail is a simple continuation of the deuterogenetic growth, and it certainly is not anything like so fundamental a change as that from proto-genesis to deuterogenesis.

Experimental Observations.

The recognition of these two growth centres in the frog is due to experimental work. It is possible by marking parts of the surface of the gastrulating egg to realize movements which other-wise escape one.

Observations by marking the surface can be made in various ways. It is possible to cauterise a few cells and the scar thus produced can be traced during a shorter or longer period of time and thus the relative movements of parts determined.

Or bristles may be inserted, at any rate in the egg of *Rana temporaria*, with many satisfactory results. A bristle makes an unmistakeable landmark, which cannot be said of the scar made by cautery. In cauterisation it is possible for the cauterised cells to slough off and to get carried away from the spot which they are intended to mark. The bristle method it may be noted cannot be used with Triton eggs. Then again stains such as Nile blue sulphate may be used, as shown by Goodale, who stained little groups of cells round the equator of the egg of *Spelerpes bilineatus* shortly before gastrulation. The effect of this treat-ment is to stain the yolk grains so that the cells go on dividing and thus the individual descendants can be followed to a certain extent. Apparently all eggs are not equally suitable for this staining method. But experiments on Spelerpes and Amblystoma were successful.

It may be said at once that the results of all these methods· of experimental research are contrary to the theory of concrescence, whether they are made upon Elasmobranch, Teleostean or Amphibian eggs.

By the bristle method it is possible in the frog's egg to make

out with considerable accuracy the area over which the main axis
of the embryo extends. It has been made quite certain that
there is a part of the embryo which is developed, so to speak,
in situ on the egg, and a part later in origin which is formed by
the activity of the blastopore lip.

These are the protogenetic and deuterogenetic regions respectively.

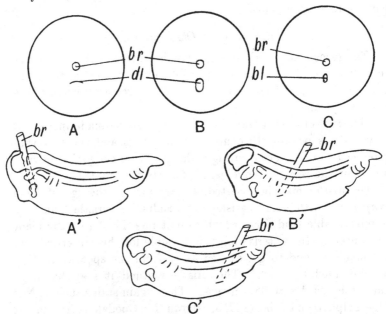

Fig. 11. *Rana temporaria.*

A, B, C, Eggs to show position of bristle; *A', B', C',* Longitudinal sections of the
same embryos after some days; *br* bristle; *bl* blastopore; *dl* dorsal lip.
Drawings of eggs and tadpoles in which bristles have been inserted to determine the parts of the embryo derived from the primary and secondary growth
centres.   In *A* the bristle was inserted 5 mm. anterior to the dorsal lip of the
blastopore at its first appearance.   In *B* and *C* the bristle was inserted about
the same distances in front of the dorsal lip but at later periods.

By carefully marking various parts at different stages of
development we are able to arrive at the following conclusion.

The protogenetic area includes:

Forebrain, eyes, mouth, heart, nasal organs, pharynx,
and probably the front end of the notochord and perhaps the
mid-brain.

The deuterogenetic area includes:

Hindbrain, spinal cord, renal organs, muscles of the trunk, and the whole tail; but probably not much of the gut except the rectum and anus.

The accompanying Fig. 11 will show the kind of evidence obtained by the method of using bristles as a means of marking regions on the egg. It is difficult to obtain the exact boundary between the influence of the two activities, because bristles cannot be placed nearer to the lip of the blastopore than about 0·5 mm. without causing serious malformations. The bristles were actually placed round the region of Brachet's blastopore virtuel (see pp. 25, 26).

Experimental Evidence as to Concrescence.

The first to test this theory by experiment was Kastschenko, who in 1888 operated upon the embryos of an Elasmobranch. He destroyed one part of the rim and found that such destruction did not prevent the formation of the main axis of the embryo.

The most complete investigation that has been made is that of F. Kopsch who performed a large number of experiments upon the developing embryo of the trout. I give here some figures showing the nature of the operation and the results.

In Fig. 12 *A* the mark *op* represents in *A* the point in the rim of the disc—i.e. lip of the blastopore, at which an injury was effected. Fig. 12 *B* shows the position of the injury at a later stage when the embryo had eight or nine pairs of protovertebrae.

According to the concrescence theory, the germinal disc rim growing round should have coalesced with the corresponding part on the other side, and formed the mid-dorsal part of the embryo, namely the neural tube, notochord, and, perhaps the proto-vertebrae. But, as the examination of sections taken through *B—b* showed, the formation of these parts is quite normal, and the injury lies far to the side—in the region of the latero-ventral part of the future body wall. The injury has brought about a failure on the part of the rim to progress over the yolk, and a corresponding *defect at the side*—which is just what one would expect if the embryo is formed in the way described.

Fig. 12 *C* is a figure of a slightly more advanced embryo

which had been operated upon in the same way as *B*. Here the
operation scar certainly appears to have moved towards the
mid-dorsal line. But if we come to examine sections, we
find that the mid-dorsal and dorso-lateral parts of the embryo
are quite unaffected. It is the part of the embryo ventral to
that which will be derived from the protovertebrae which is
injured.

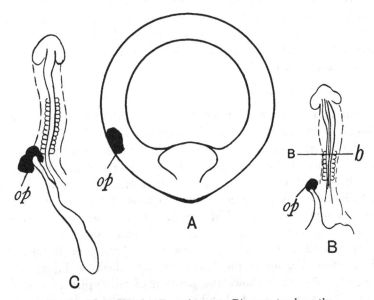

Fig. 12 after Kopsch. Egg of trout. Diagram to show the
result of injury to the rim of the blastoderm.

A blastoderm; *op* injury; *B* the embryo some days later; *C* another specimen.

The notochord, neural tube, and even the protovertebrae are
normally developed, but a little of the tissue beyond—that is to
say, ventral to this point—is lost. This means that it is the *side*
or lateral body wall of the future animal which will show a
deficiency, and this is only what one would expect upon the non-
concrescence theory.

An injury made in the mid-dorsal line completely prevents
any addition to the mid-dorsal region of the embryo. If the
embryo were formed by concrescence this would not be so.

An injury inflicted at a point between that of Fig. 12 *A* and the tail bud brings about a defect in the dorso-lateral region, leaving intact the median structures, such as the notochord and the floor of the neural tube.

These experiments, to my mind, are quite conclusive against the theory of concrescence or confluence in any form, but it is only fair to say that this view is not held by all.

Even Kopsch, from whose work I have taken these figures, believes that his experiments indicate a certain amount of concrescence for the hinder end. He shows how, for instance, as in Fig. 12 *C*, that there is a movement of the point of injury nearer to the median line. It is true there is a change of position with reference to the yolk mass, but where is there any evidence that the dorsal or dorso-lateral region is affected by such an injury? It is in the much distended latero-ventral region of the young embryo that the injury has been made, and it is in that region of the fully-formed embryo that the injury persists. So long as the notochord and the floor of the neural tube are uninjured by such experiments as these, so long is there a complete absence of evidence for concrescence or confluence.

Goodale stained groups of cells in the equatorial region just prior to gastrulation. These groups lengthened out meridionally and converged upon the blastopore.

This showed that as deuterogenesis effected its work, the lips of the blastopore progressed leaving trails behind them which showed not a trace of concrescence.

LECTURE II

We have been able to come to a fairly decisive conclusion with respect to the growth in length of embryos of the Anamniate Vertebrates. We have arrived at this conclusion chiefly through anatomical observations and also from evidence obtained by actual experiment upon certain fishes and Amphibia.

We have seen that the production of undifferentiated cells which go to build up the general form of the embryo occurs around two centres of activity. The first may be said to be around the protoplasmic centre of the fertilised egg, and to begin with segmentation and to end (so to speak) with the accomplishment of gastrulation—this is the primary or protogenetic growth centre. The second begins with the formation of the blastopore lip at the commencement of gastrulation, and may be said to be around a centre situated in the middle of the blastopore, for the form which this centre of growth takes in the Anamnia is annular.

This is the secondary or deuterogenetic growth centre. It is placed eccentrically to the primary growth centre and the effect of its action is the growth in length of the embryo, i.e. deuterogenesis. New tissue is added on to all previously existing tissues with which it is in contact.

Hence we are able to distinguish between protogenetic ectoderm and deuterogenetic ectoderm, protogenetic part of the nerve tube, and deuterogenetic part of the nerve tube, protogenetic part of notochord, deuterogenetic part of notochord, and so on.

The clear appreciation of this fact renders intelligible the differences said to occur in the method of formation of the notochord and explains how it can arise either from endoderm or mesoderm or ectoderm. It does away also altogether with the idea of gastrulation occurring in two phases as Hubrecht, Keibel and some other embryologists have asserted, because the two phases are quite distinct, the first being the real gastrulation, the effect of protogenesis, while the second so-called phase of gastrulation is a totally different phenomenon, namely the growth in length of the already formed gastrula by accession of material from behind, so that a spherical form is converted into a cylinder.

In all the Anamnia the rule is that the gastrula cavity or archenteron is formed with an opening to the exterior—the blastopore. In all the Anamnia the form of the deuterogenetic proliferating area tends to be annular and the blastopore is the hole in this ring.

Conditions in Amniota.

In the Amniota we find a very different state of affairs.

It is extremely doubtful if the part of the gut cavity which I have termed archenteron (Fig. 13), as distinct from metenteron ever forms in connection with a blastopore.

An opening between it and the exterior may occur later but that is of a different character and it is really not an archenteric, but a metenteric opening. These two openings are what Brachet calls blastopore virtuel and blastopore réel, Fig. 13.

The metenteric opening or notopore of some authors (blastopore réel) (of which the more dorsal part becomes neurenteric canal for a while) subsequently disappears, and the ventral part becomes the anus.

In the Amniota the archenteron always arises as a split either between cells, or else as vacuoles among cells which will be endoderm cells; and in most cases it does not arise so much as the result of the geometrical effect of segmentation (as in Amphioxus or the Frog), Fig. 6, as by internal secretion of fluid, Fig. 15 C.

There is only one exception to this, and this a very obvious exception indeed, namely, the segmentation of the *Marsupial* mammals. But the process in this case is very different from that of any other vertebrate, though at the same time it may be said that it can be derived fairly easily from a meroblastic type of egg similar to that of Prototheria or Sauropsida.

In Anamnia the archenteron is formed at the same time that the deuterogenetic centre begins to come into existence.

In Amniota the archenteron is formed before the deuterogenetic centre has come into existence.

In Anamnia the archenteron is formed either by invagination of one wall of the blastula within the other, or by a process of splitting or delamination, producing a split which is always open to the exterior, or by a combination of both processes. In

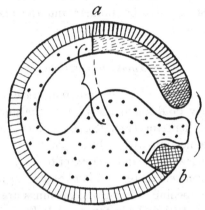

Fig. 13.   Diagram to show the opening of the blastopore in Anamnia—the line
    a—b divides protogenetic tissue (left) from deuterogenetic (right).
Dots and fine lines protogenetic endoderm and ectoderm respectively; broken
    lines = deuterogenetic endoderm; thick lines = deuterogenetic ectoderm; the
    bracket on the right marks the metenteric opening, or notopore (Brachet's
    "blastopore réel"), the bracket to the left marks the archenteric opening
    ("blastopore virtuel").

Amniota the archenteron is formed by internal secretion of a cavity among or within the yolk cells.

These differences are in correlation with a physiological process which differentiates the mode of development very sharply indeed from that of the Anamnia. This is the fact that in an early stage of development, a fluid is secreted into the two great cavities of the embryo, namely the *archenteron* and the *coelom*.

This condition of oedema may be said to be the physiological essential cause of the formation of the amnion and allantois, which are the most distinguishing features of the Amniota as compared with other vertebrates. If this is so we can see why there should not be a blastopore during the time of the formation of the archenteron. If there were a blastopore at this time, clearly there could be no cavity produced by swelling.

The amnion is essentially an organ of the coelom. This view has been taken by Hubrecht in his paper upon the "Early development of the Hedgehog and its significance for Phylogeny of the Amnion."

But it has also probably arisen in correlation with a distended archenteron.

In Elasmobranchs there is a very heavily yolked egg, but there is no amnion. Nevertheless it is very interesting to note in the Elasmobranchs a condition which might very easily lead to the

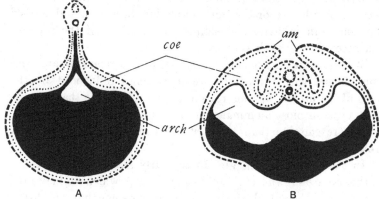

Fig. 14. *A*, Scyllium; *B*, Amniote.

Diagrammatic section. Black = endoderm; dotted line = mesoderm; line barred with white = ectoderm; *coe* coelom; *am* amnion; *arch* archenteron.

formation of an amnion of the amniotan type. In Dogfish embryos of about half an inch in length, at a time just after the closure of the blastopore, there will be found two large coelomic cavities upon the yolk sac, that is to say cavities of the so-called exembryonic coelom, which are morphologically and physiologically in the position occupied by the lateral folds of the amnion of an amniote (Fig. 14).

In Scyllium there is no accumulation of fluid within the gut cavity, and the rather long yolk stalk keeps the so-called embryo part well away from the yolk sac.

But a very slight alteration of conditions could convert this into an amniote as Fig. 14 illustrates.

There is one other point to be insisted upon in considering this question of growth in length in Amniota—not only is there this fluid enclosed in the archenteron, but there is an increasing amount of pressure, so that we have here to deal not only with a hydrostatical condition but also with a hydrodynamical condition, and this must have its effect upon the mechanics of development.

The development of the Rabbit.

We will now take the formation of growth centres in the embryo of the Rabbit in more detail. This animal is chosen because of all vertebrates none other has been so far described that shows in so distinct a manner the relations of the primary and secondary growth centres.

We know that the blastocyst of the Rabbit attains a large size from causes within itself before it becomes influenced by the walls of the uterus, for this reason it is more suitable than are the blastocysts of most mammals, and because of the absence of yolk it is more suitable than those of birds or reptiles, for study of growth centres.

In the accompanying Fig. 15 the early stages in an ordinary Eutherian are shown—such as that of the Pig, which we may take as typical of the sub-class—in which there is some, but very little, indication of what has been called inversion (entypy).

The chief points are that the egg has but little yolk though some is present in the form of fat drops. The segmentation is holoblastic, the four segment stage is criss-cross, Fig. 15 *A*, and the result of segmentation is a solid sphere of cells all much of the same size—this stage is termed the morula, Fig. 15 *B*. A cavity arises by the formation of vacuoles within some of the more internal cells, Fig. 15 *C*. The vacuoles run together, the fluid increases and the cavity gets larger and larger till it occupies a position that leaves only a heap of cells at one pole, and a layer reduced to a single cell in thickness elsewhere, Fig. 15 *D*.

The heap of cells becomes flattened, and three layers are distinguishable at this point; the outer is termed trophoblast, the middle ectoderm, the inner endoderm, Fig. 15 *E*.

The middle layer or ectoderm is distinguishable earlier, and in some mammals very much earlier, than the other two are dis-

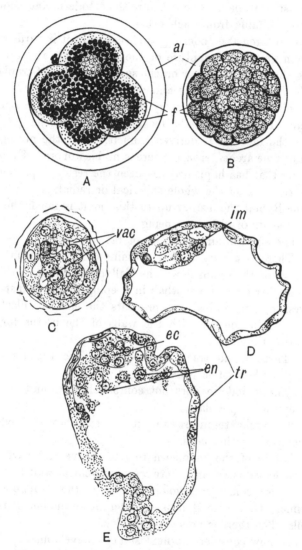

Fig. 15. Pig.

A, 4 segment stage; B, morula stage; C, blastocyst showing vacuoles; D, 4½ day blastocyst; E, 7 day blastocyst; al albumen layer; ec ectoderm; en endoderm; f oil drops; vac vacuoles; tr trophoblast; im inner mass.

tinguishable from each other. In some mammals it may be different, as in Manis and Galeopithecus, where it is said that the trophoblast is to be made out before the ectoderm and endoderm are distinguishable from each other.

Then the circular ectoderm plate becomes bent inwards and by its growth causes the rupture of the trophoblast above it, thus allowing the plate to appear on the surface; pieces of trophoblast upon the surface degenerate and disappear.

All this time the symmetry has been markedly radial or spherical. The cavity of the blastocyst becomes the gut, or at any rate the fore gut is derived from it. It is the protogenetic enteron or true archenteron. There is no blastopore. Up to now everything that has happened has been the result of protogenesis and the result is on the whole spherical or radial.

In the Rabbit the history up to this point is much the same, the chief points of difference being:

(i) The egg contains protein not fat.

(ii) There is a very tough albuminous layer laid on during the passage of the ovum down the Fallopian tube, and this acts like the leather coat of a football in keeping the more distensible inner vesicle taut, and also it prevents the embryo (blastocyst) from coming in contact with the walls of the uterus for some considerable time.

(iii) There is no trace of inversion. The rupture of the overlying trophoblast is general.

(iv) The endoderm does not completely surround the inner surface of the trophoblast.

(v) The archenteron arises more as a split between cells than as vacuolations within cells.

The failure of the endoderm to line the ventral part of the blastocyst looks as though there were no ventral wall to the gut cavity if—as some embryologists think—the trophoblast is ectodermic. Of course if it is endodermic or homologous with yolk cells, then there is no such difficulty.

We can now consider the next stage in development. Protogenesis has come to the end of its solitary reign. The radial symmetry to which it gave rise now becomes profoundly modified by the origin of a secondary growth centre resulting in growth in length. The conditions in the Rabbit are such that the plane of the deuterogenetic area coincides at first with that of the protogenetic area, instead of being at right angles to it as in the Anamnia.

This is (i) partly on account of the great distension of the archenteric space, and (ii) partly because the deuterogenetic centre first shows itself as a circular spot instead of a ring.

In the Rabbit we have the stages of development of the two centres shown in quite a diagrammatic form in Fig. 16.

During the seventh day the circular embryonic area is marked out very distinctly in front, but less distinctly behind.

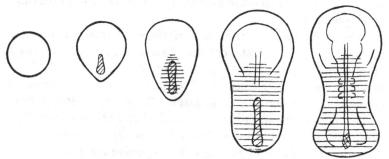

Fig. 16. Rabbit seventh and eighth day. Diagram
to show the secondary growth centre.
The white areas are protogenetic, the shaded areas are deuterogenetic, the doubly
shaded portions represent the primitive streak.

The embryonal area is coterminous as yet with the ectoderm. The surrounding superficial cells are all trophoblastic. Internally the endoderm lines both the trophoblast and ectoderm. About the time that the deuterogenetic centre begins to appear, there is a slight thickening of the endoderm just under the anterior and lateral margins of the ectoderm. This is the first sign of any mesodermal formation, but no cells are actually budded off here until the deuterogenetic tissues have become apparent. This semicircle is however the first sign of the mesoderm from which the heart and pericardium will be developed, but we will confine ourselves at present to the external characters, that is to say to the general form. The deuterogenetic centre begins as a proliferation of cells of the outer layer of the hind part, probably of the actual edge of the embryonal area.

The proliferation is from the under surface. But we must presume also an addition of cellular units to all tissues in contact with the proliferating area. The cells which are proliferated within give rise chiefly to mesoderm and anteriorly also to notochord, that is to say deuterogenetic notochord. We will return to this point later.

There can be no doubt that here is an area of intense cell divisional activity.

This centre of activity has apparently arisen suddenly. It is close to but not coincident with the centre of the primary centre, which we may take as being practically in the centre of the circular embryonal area.

Now what must be the effect of the counter action of these two centres? We must consider them for the moment apart altogether from any phylogenetic point of view. Here we have two actively productive centres of tissue, one of which is especially active. They lie on the same plane. They are subjected from within to an ever increasing hydrostatical and therefore a hydrodynamical pressure. Surely the result must be that they tend to elongate along the lines joining their respective centres.

If one area is more yielding than the other then that area will be elongated the most.

What are the conditions here? Inside there is an increasing hydrodynamic agent. This is shown by the way in which the albumen layer is kept taut and becomes thinner, and later it is seen by the expansion of the uterine wall. Therefore the cells are stretched also. But where there is also a tendency to expansion due to cell multiplication—there will be an additional tendency for the cells to stretch.

Again if the cell multiplication is greater in one area than in the other, the area in which the cell multiplication is greatest will give way more easily than the other.

The deuterogenetic area seems to be an area in which cell multiplication is very intense, and therefore we find that this part expands more markedly than the protogenetic area.

The result is that the posterior half of the embryonal area is drawn out, but the anterior half of the deuterogenetic area is drawn out to a still greater extent, the latter appearing as a conical

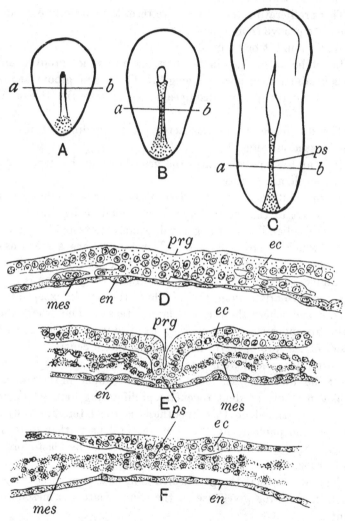

Fig. 17. Rabbit embryonal areas.

D is a transverse section through *A*, 172 hours; *E* through *B*, 180 hours; *F* through *C*, 192 hours; in *A* the primitive streak is elongated and slightly grooved; in *B* the primitive streak is at the maximum length and the groove at maximum depth; in *C* the groove has disappeared. *en* endoderm; *ec* ectoderm; *mes* mesoderm; *ps* primitive streak; *prg* primitive groove.

or wedge-like strand of thickened proliferating tissue running into the also drawn out but less conspicuous hinder region of the protogenetic embryonal area.

This proliferating area is what we term the "primitive streak." Along it a groove soon appears.

Why should it be grooved?

Let it be noted that in the first place it is not grooved until it has attained a considerable length. There is no groove at first.

Secondly the groove disappears long before the streak disappears.

The difference in the characters at these different times is seen in sections, Fig. 17.

It is at least possible that the grooving may be very largely due to mechanical causes.

As the deuterogenetic mesoderm passes to either side by being crowded out in the median line, it will tend to be drawn away from the median line by the general expansion of the blastodermic vesicle which is still going on. This must place some tension upon the under surface of the proliferating area, if there is any continuity, so that each side will be drawn away. The result is that a groove arises along the middle of the streak, except at the anterior end where there is less lateral tension, and where there is also an opposing compression in front, due to the antagonistic effects of the protogenetic and deuterogenetic centres.

Posteriorly also there is no groove or if there were any it would tend to be transverse, as indeed it is actually found in the Sparrow.

The part of the deuterogenetic proliferating area which will be most expanded on this hypothesis is the latero-anterior, or, speaking morphologically, the dorso-lateral part, that is to say the part equivalent to the dorso-lateral part of the blastopore lip of the Frog, Fig. 18.

Thus we see that the part which is most expanded is the part which in the Frog gives rise to the neural plate and mesodermic somites.

If we look at figures of the Rabbit's blastoderm from the time the primitive streak first appears until the time of the formation of the first 7–8 protovertebrae—we can see throughout the following features very well defined.

(i) The general radial character of the protogenetic area,

(ii) After a certain moment the primitive streak becomes contracted—shortened and thickened,

(iii) An additional part of the embryo has been intercalated between the primitive streak and the radial protogenetic area.

This intercalated region is approximately all that part of the embryo which includes the protovertebrae. Now we find that in the Amphibian (Frog) it is just that part which is formed from the lips of the blastopore.

If we can assume that this intercalated region has been added by, and at the expense of, the primitive streak, we arrive at the

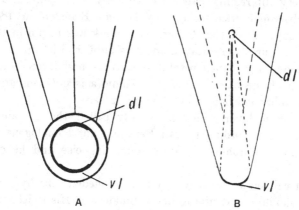

A   B

Fig. 18. *A*, Frog; *B*, Rabbit.
Corresponding deuterogenetic proliferating areas; *vl* ventral;
dl dorsal lip of blastopore.

conclusion that physiologically the primitive streak is identical with the blastopore lip region of the Anamnia.

To prove this by experiment is probably impossible in Mammals, but in Birds the supposition has been put to a practical test.

If we can assume that the primitive streak of the Bird is the same as that of the Rabbit, then we can say that experiment proves that the primitive streak is converted into the meta-merically segmented part of the animal, but that the anterior part of the animal is developed quite independently of the primitive streak.

Thus again we find that in the Mammal and Bird, there is a protogenetic and a deuterogenetic region; and that they correspond to one another.

The protogenetic region gives rise to the fore brain, the mouth, the pharynx, the heart and pericardium, and the deuterogenetic region to all the rest of the animal in both cases.

Fate of the Primitive Streak.

We have so far considered the question of the fate of the primitive streak generally. We must now take it more in detail.

A very interesting paper has recently been published post-humously which was written by E. van Beneden in 1889 on the development of the Rabbit. The work was begun in 1881–2, when the figures were drawn, and was published in 1912.

The view van Beneden takes is that the first formed didermic condition of the blastocyst of the Amniota is not comparable to the didermic condition of lower vertebrates.

The inner layer which is usually called the endoderm or hypoblast is, according to van Beneden, not homologous to the endoderm of Amphioxus. It is something else and he calls it *Lecithophore.*

As I will shew presently, what van Beneden calls hypoblast is what I should call deuterogenetic endoderm. His lecithophore is what I term protogenetic endoderm and this, according to my way of looking at things, is very clearly comparable to the true endoderm of the lower chordates.

Further, van Beneden takes a very different view of the development of the Rabbit from that outlined above. He takes no heed of internal hydrodynamic pressure, or tensions, or anything of that kind. To him the cells at the lower pole of an expanding blastodermic vesicle of the Rabbit, are squamous not because of the pressure from within to which they are subjected—but he thinks on the contrary that the blastodermic vesicle expands because the cells stretch themselves out.

He does not regard the cavity of the vesicle as archenteron, nor the inner layer of cells as endoderm and says that "Gastrulation" does not occur until after the primitive streak has been formed, and also that the canal of Lieberkühn is the invagination cavity and is therefore archenteron. In his view the blastocyst of a six-day Rabbit before the primitive streak begins to form is not a gastrula, but a blastula, Fig. 19. The

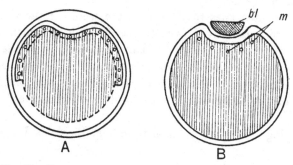

Fig. 19.   Diagrams to illustrate van Beneden's interpretation.
A, Mammal; *B*, Elasmobranch.   *bl* blastodisc; *m* merocytes; the lined area is the lecithophore.

inner layer of cells together with the fluid van Beneden calls *lecithophore*, and he says they are homologous to the unsegmented yolk of an Elasmobranch with its merocytes.

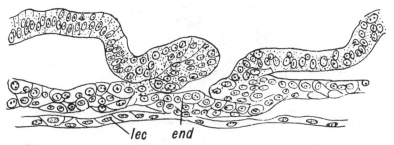

Fig. 20 after van Beneden.   Rabbit eight days.   Plate XIV, Section 14.
Transverse section through embryonal area.   *end* (endoderm) archenteric plate of van Beneden; *lec* lecithophore.

Clearly if Rückert is right in ascribing the origin of the merocytes to the accessory spermatozoa, this homology will not hold good, but it might apply to the Teleostean egg.

Gastrulation according to van Beneden occurs by invagination, producing the groove along the primitive streak and the canal of

Lieberkühn. This area which "invaginates" is said to be endodermal and since it clearly does not invaginate at the posterior
end, for there is no groove there—it constitutes the equivalent of
a yolk plug. Therefore after invagination we have the relation
as shown in Fig. 20.

Van Beneden tries to show that the whole of the notochord
and gut lining is derived from this archenteric rudiment and he
brings forward some very remarkable evidence from the corresponding stages in the Bat in support of this. But even with
sections which favour his views there still remains a considerable
amount of gut lining which must be derived from the "lecithophore."

Thus Fig. 21 which is taken from Fig. 15 of his Plate XXI.

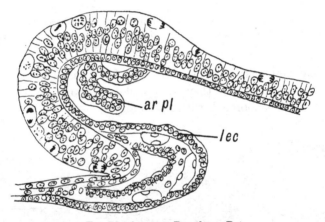

Fig. 21 after van Beneden. Bat.
Sagittal section through the anterior end of an embryo with six protovertebrae.
ar pl remains of archenteric plate projecting into the foregut; *lec* lecithophore.

It is however not at all certain that the remnant of archenteric
plate has not shifted its position. It is degenerating tissue and
may quite well move, as I have seen degenerating trophoblast
move in the Pig.

Again the pericardium is very clearly indicated at quite an
early stage by the thickening of the inner layer (lecithophore of
van Beneden) and this layer certainly occupies the position
marked *lec* in the above figure.

Van Beneden regards the whole of the notochord, the whole of the gut lining and the whole of the mesoderm, with the possible exception of the endothelial lining of blood vessels, as being derived from the primitive streak, i.e. by invagination. The primitive streak is converted into the embryo by *concrescence of its lips*. Even if the primitive streak does become converted into "embryo" by concrescence, this does not necessarily prove that the original blastopore existed along the dorsal surface of the ancestral chordate.

Because:

(i) The primitive streak appears in the first instance as an already closed blastopore.

(ii) The grooving is I believe due to mechanical causes.

I would interpret the bat condition (Fig. 21) as follows.

All van Beneden's archenteric plate I regard as deuterogenetic. His discarded "lecithophore" is protogenetic. His figures of bats certainly do not support his contention that no notochord is formed from the lecithophore, (v. Plate XVIII, fig. 7, 8). There is evident thickening along the middle line due to proliferation of nuclei at right angles to the surface.

Gastrulation and Growth in Length in Birds.

The segmented part of the ovum at the end of seven hours after fertilisation consists of a little cap of cells occupying the centre of the upper pole. These cells have been produced by the segmentation of the germinal disc. They may be said to be in cellular continuity with the so-called unsegmented yolk mass around and below them. We may as well term this cap of cells the blastoderm. It is important to remember that this cap of cells is continually being added to peripherally and perhaps beneath as well. It should be noted that whenever a new segment is cut off from the peripheral yolk mass the formation of such a new cell is preceded by the division of a nucleus, and while one daughter nucleus is contained in the new cell thus cut off and added to the blastoderm, the other daughter nucleus remains behind in the yolk mass, and after awhile it again divides with the result that another cell is added to the blastoderm, and so on. Therefore morphologically we must look upon the "yolk mass" as being merely the more ventral segment of a fully segmented ovum.

If we examine the egg of a sparrow just after it has been laid it has in many cases very much the appearance just described. There is however usually a slight cavity below the lowest layer of small cells and the general yolk mass. In the case of the new laid egg of *Gallus* the cavity is much more evident.

The blastoderm consists of a definite outer membrane of cells, of practically a single layer in thickness in its central part though many layered round the peripheral area. A distinct cavity, Fig. 23 *sg*, filled with fluid and containing many more loosely arranged cells underlies the thin central part of the outer membrane. In surface view this central part appears dark owing to the translucency of the membrane itself and the cavity into which we are looking; just as a window appears dark when looked into from outside. The thickened peripheral part, as it is opaque and rests on a solid yolk, looks lighter in a surface view, Fig. 22. These are the areas of the blastoderm known as area pellucida and area opaca respectively. In section one would always see in addition to the cellular membrane, the non-cellular *vitelline* membrane extending over blastoderm and yolk alike.

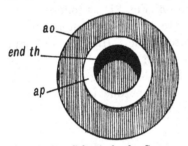

Fig. 22 after Schauinsland. Sparrow egg new-laid.

ap area pellucida; *ao* area opaca; *end th* thickening of endoderm.

The cavity containing the fluid lying under the area pellucida, is usually called the subgerminal cavity, and it continually increases in size by further accumulation of fluid as development proceeds. It is in fact the future gut cavity, and ought more strictly to be called the archenteron. Until recently it has generally been assumed that it arises in the bird's egg just as the corresponding cavity arises in the reptile's egg, by the gradual accumulating of fluid between the small cells of the segmented blastoderm and the large cells forming the unsegmented yolk.

Then as the result of the gradual multiplication of the cells which lie between the outer layer and the yolk, and their arrangement into a definite membrane, an inner sheet of tissue is produced, forming eventually the gut epithelium.

This, at any rate, is the condition attained after a few hours of incubation and as soon as this has been accomplished we can recognise the presence of what are known as the primary germinal layers, and the process of *gastrulation* has been finished. Gastrulation is simply the formation of a cavity which is the future gut cavity, and the lining of this cavity by a definite layer of cells, the future epithelium of the true gut or mesenteron.

There is as yet, Fig. 23, no opening to the exterior, no mouth nor anus.

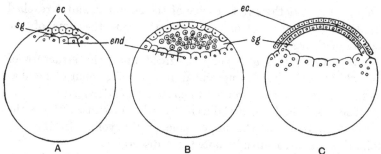

Fig. 23. *A*, Bird in oviduct; *B*, bird laid; *C*, bird after short incubation. *ec* ectoderm; *end* endoderm; *sg* segmentation cavity.

The outer membrane about this period becomes dissociated from the underlying layer, at any rate at its margin, and it extends beyond it and creeps slowly over the yolk.

The names of these several layers and parts are as follows.

The outer layer is called the ectoderm, the inner the endoderm.

The endoderm is obviously continuous at its margin with the general yolk mass, and is greatly thickened here.

The thickened margin is known as the *germinal wall*, the cavity as the *subgerminal cavity*, or archenteron.

If we take the account just given of the origin of the gut cavity in the bird's egg as correct, it will fit in very well with the origin of the gut cavity in both the other great classes of Amniotes, the Reptiles and Mammals.

The condition of a solid morula is converted into the condition represented in Fig. 24, by accumulation of fluid in spaces among the lower layers of the segmented germinal disc, or between them and the largest yolk segments forming the basal part of the ovum.

An entirely different account of gastrulation in the Pigeon's egg has been given by Patterson, which if confirmed would necessitate an entirely different interpretation of the facts so far described. But it is extremely difficult to reconcile his account of the early development of the bird's egg with the facts of Mammalian and Reptilian embryology, and I am very doubtful as to whether Patterson's conclusions can be accepted (v. my paper on "Gastrulation in Birds," *Q. J. M. S.* Aug. 1912).

The chief point in dispute concerns the exact mode of formation of the lower layer or endoderm. That we must leave as uncertain, only noting that the anterior edge of the germinal wall is reached by the advancing edge of endoderm at about the time of the laying of the egg.

All investigators however are agreed upon the structure of the blastoderm after laying and during the first hour of incubation, and we may now proceed to consider its later history.

The freshly laid egg of Gallus if it has been fertilised, shows the blastoderm lying on the upper pole of the yolk. In this the following parts as already noted are discernible:

Area pellucida (through which the nucleus of Pander can often be seen);

Area opaca.

Surrounding the margin of the area opaca a slight groove usually runs separating the blastoderm from the surrounding yolk.

Sections show that this blastoderm consists of a segmented disc lying loosely on and connected with the sub-lying yolk, and having an outer well-defined layer of cells, and under this a large number of loosely arranged cells, which form an incomplete roof to a distinct cavity between them and the yolk. This is the *subgerminal cavity*. There is a good deal of variation in the state of the egg when laid, sometimes it is further advanced than this. Also some blastoderms may be broader, measuring quite a millimetre more than others. In such cases there is a greater attenuation of the cellular layers, and the subgerminal cavity is much larger.

The looser inner layer of cells when the section is taken along the direction of the future main axis, is thicker towards its posterior end.

Of course after laying there is an almost complete cessation of the developmental processes as the egg cools, so the egg remains in a dormant state, though in hot summer weather development makes some slight progress. Then as soon as incubation begins the cells again become active and the changes effected are rapid. During the first few hours the chief change is in connection with the margin of the blastoderm.

The groove seen in the unincubated egg quite disappears and the outermost layer of cells extends over the yolk beyond the zone where nuclei lie scattered within the yolk. This layer is now quite distinct from the underlying cells throughout the whole area

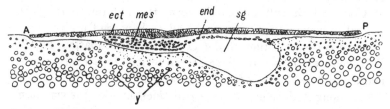

Fig. 24 after Duval. Section through blastoderm of chick
incubated ten hours.
A, anterior; *P*, posterior. *ect* ectoderm; *end* endoderm; *mes* mesoderm;
sg segmentation cavity; *y* yolk (subgerminal).

and it constitutes the *ectoderm*. Its progress in the future over the yolk is by its own interstitial growth. It receives no additional cells or nuclei from the yolk.

The loose cells below have by the 10th hour of incubation, Fig. 24, formed a continuous sheet of tissue, this is the endoderm and it is connected all round its margin with the yolk, which there contains many nuclei, and it is called the germinal wall. According to some authors, Duval, Marshall and Balfour, some of the cells of this loose mass are left between the sheet of endoderm and the ectoderm, and they are said to be more numerous in the anterior part of the area pellucida. On the other hand other authors (e.g. Lillie) deny the presence of these cells.

There can be no doubt about their presence in the Sparrow, where they form a many layered mass circular in outline in the anterior part of the blastoderm. These are the first traces of the middle layer of cells or *mesoderm*.

The ectoderm even at the time of laying is thicker near the centre of the area pellucida than elsewhere. In Gallus after some 5–10 hours of incubation this thickening is more marked still over a wide area occupying the more anterior part of the area pellucida, and a little later still a much more decided thickening can be seen in the more posterior part of the area pellucida extending from about the centre of the area backwards nearly to the margin of the posterior part of the area opaca. This thickening is the first sign of a structure which has long had the name of primitive streak, but it should be recognised at once as the growing point for the increase of length of the embryo, while the less well-defined thickening of the ectoderm in front is the beginning of the plate of tissue which will ultimately form the anterior part of the central nervous system—the fore and perhaps mid brains.

Such is the condition at the 11th hour of incubation, Fig. 25. Transverse sections through these two regions reveal certain well-marked differences. A transverse section through the anterior region along *A–B* shows us a decided thickening of the ectoderm, but it is a wide diffuse thickening, having a quite sharp inner surface; a few scattered cells are present between the layers which are the mesoderm cells already referred to: while a section along *C–D* shows a much more restricted area of thickening which has a ragged inner surface from which mesoderm cells are being actively proliferated.

Up till now in surface view the whole blastoderm has appeared circular, as also the area opaca and the area pellucida. But now the outline of the area pellucida becomes elongated and it can be clearly seen that this elongation of the area pellucida is an elongation backwards, and that this elongation involves the elongation of the primitive streak. By the 18th hour the primitive streak has stretched backwards and it now reaches from the centre of the original circular area almost to the area opaca. It is a more or less straight line, and is deeply grooved along its whole length except at the two ends. The groove is deepest at the anterior end, and it ceases abruptly against a rather well-marked mass of cells which is sometimes called "Hensen's knot" or the primitive knot, but posteriorly it gradually shallows out into a rather expanded thickening which forms the end of the streak.

In front of the primitive streak the wide thickened area of the ectoderm is now more sharply marked out from the rest of the ectoderm. It constitutes the neural plate and is traversed longi-

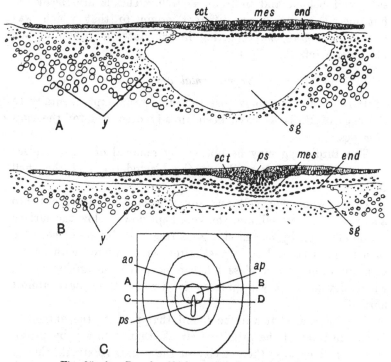

Fig. 25 after Duval. Chick at 11th hour of incubation.

C, surface view; *A* and *B*, sections taken along the lines *A—B* and *C—D* respectively. *ect* ectoderm; *mes* mesoderm; *end* endoderm; *sg* subgerminal cavity; *y* yolk; *ps* primitive streak; *ap* area pellucida; *ao* area opaca.

tudinally by a rather wide shallow groove, the *neural groove*. This groove is absent just in front so that posteriorly the neural plate is partly divided by the shallow neural groove, but in front the two halves are united.

The study of a series of transverse sections makes the features peculiar to all these several parts clear and also explains their relations to the whole egg and to each other.

The most important feature is perhaps one which it is difficult to see in the chick in surface view.

This is the fact that the primitive streak is clearly an area of intense cell proliferation and that cells are being continually budded off from its whole length into the space between the ectoderm and endoderm; and that these are spreading out to the sides and behind and in front as well. This is also *mesoderm* but mesoderm having a different origin to that which was mentioned before, as occurring in the anterior region of the blastoderm, cf. Fig. 25 *A*.

Experimental proof.

It is possible to apply just the same tests experimentally to the eggs of Birds as have been already described for the eggs of Frogs.

The Bird's egg may be opened by removal of a small piece of the shell, a mark made on the blastoderm, and the shell replaced.

Or the whole egg, yolk and albumen, may be turned out into a suitably sized vessel, and the yolk kept down below the surface of the albumen by means of a glass ring; the bristle or hair can then be inserted and the vessel, after being covered with parchment or even with a glass lid, can be placed in an incubator, where development will proceed for two or three days almost normally.

By this means it will be found that part of the animal is formed in front of the primitive streak, that is to say by protogenesis, and part by the activity of the primitive streak, that is to say by deuterogenesis, and that the two areas correspond to the two parts derived from those centres in the Amphibian.

LECTURE III

The conditions in the Reptilia are in some respects intermediate between those of the *Amphibia, Birds* and *Mammals*. In each case there is no early formed blastopore and the archenteron is formed by the accumulation or secretion of fluid among the yolk cells, that is to say among the endoderm cells.

This is very clearly shown in Will's figures of Platydactylus. There are the same primary and secondary growth centres at work, but the conditions under which they work are quite different. The difference in the conditions which determine the difference in effects seems to be (i) the less abundant, and less rapid accumulation of fluid or perhaps it would be more accurate to say the less early accumulation or secretion of the fluid among the endoderm or yolk cells; (ii) the form of the secondary "centre" is discoid instead of annular. The condition is intermediate between that of an Amphibian like Hypogeophis, and a Mammal, as Fig. 26 will illustrate.

There would seem to be an accumulation of fluid among the endoderm cells in Hypogeophis, but this never causes a swelling because the space wherein it accumulates opens at once to the exterior by the "blastopore" shown in Fig. 27.

In the Bird and the Mammal, as we have seen, the fluid accumulates very rapidly and there is no way for its escape, so the cavity enlarges, becoming the subgerminal cavity in the one case, and the blastocyst cavity in the other. In both cases it eventually forms the anterior part of the gut cavity. It can therefore be called the archenteron.

In the Reptile there is an undoubted accumulation of fluid as in the Mammals and Birds and there is at this time no blastopore opening—but there is a difference and this is a difference of degree. In the Reptile the accumulation is slower.

If we compare the stages of Mammal, Bird and Reptile at the moment when the deuterogenetic centre becomes evident we find

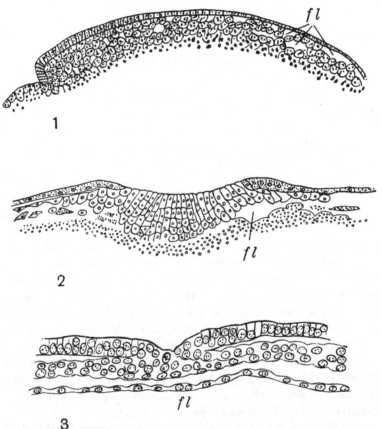

Fig. 26. 1. Hypogeophis after Brauer, 1907.   2. Platydactylus after Will, 1892, showing little fluid.   3. Vespertilio after van Beneden, 1912, showing much fluid. *fl* fluid.

very marked differences. In the Mammal the blastocyst cavity is already large and is rapidly growing by continued increase of hydrodynamical pressure producing tension in the layers above. In the Bird the same condition is almost as evident. It is only masked by the presence of the yolk mass, Fig. 28.

But in the Reptile at this stage there has been very little accumulation and therefore presumably very little tension.

The accumulation of fluid does occur later, but not until the deuterogenetic centre has become well established, and the result

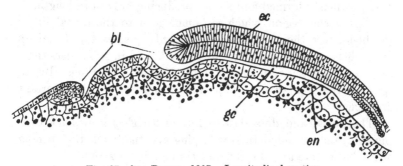

Fig. 27 after Brauer, 1907.   Longitudinal section
through the egg of Hypogeophis.
bl blastopore; *ec* ectoderm; *en* endoderm; *gc* future gut cavity.

is therefore a mass of cells in the form of a heap instead of a long drawn out streak, as in the case of Mammals and Birds.

When the accumulation does come, that is to say when the stretching is produced which results in the rupture of the posterior

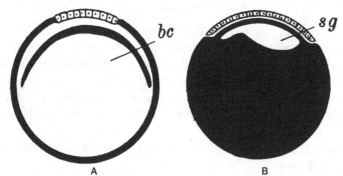

Fig. 28.   *A*, Mammal;   *B*, Bird.
Diagram to compare corresponding stages in the formation of the segmentation cavity. *bc* blastocyst cavity; *sg* segmentation cavity.

part of the roof of the subgerminal cavity or archenteron, there occurs at the same time a perforation within the deuterogenetic tissue which corresponds in position with the metenteric opening referred to in a previous paragraph.

At the same time the deuterogenetic centre has been acting just as it does in an Amphibian; that is to say dorsally it has been adding on new tissue to the previously existing tissue, new ectoderm to the primarily formed ectoderm, new neural plate to that previously formed and so on, producing growth in length.

This dorsal region which is homologous to the dorsal lip of the blastopore therefore grows backwards, just as the dorsal lip of the blastopore of the Frog grows back, and in some cases there results a condition very similar to that found in the Frog. But in spite of the similarity in appearance there is an all-important difference.

The dorsal and dorso-lateral lips *in the Frog* grow back over the yolk, and an endodermal yolk plug eventually fills the opening of the blastopore and projects out through it for a time.

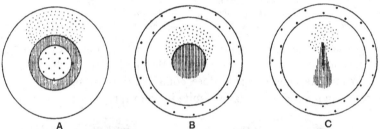

Fig. 29.   *A*, Amphibian; *B*, Reptile; *C*, Bird or Mammal.
Small dots, protogenetic tissue; large dots, yolk; lines, deuterogenetic tissue.

In the Reptile, the dorsal lip, and the dorso-lateral lips are growing back over the morphologically coalesced ventro-lateral and ventral lips of the blastopore—not over yolk or endoderm at all—that is to say, not over protogenetic tissue but over deutero-genetic tissue.

There is no blastopore in the Reptile, and therefore there can be no yolk plug.

The canal formed in this way is neurenteric canal as Kupffer correctly named it. It has also been termed mesodermal sac by Hertwig. It is wholly deuterogenetic.

The accompanying diagram, Fig. 29, should make this clear as seen in surface view. The small dots, large dots and shaded lines represent the protogenetic tissues, yolk and deuterogenetic tissues respectively.

Growth in length occurs by the activity of the deuterogenetic mass, which as in the Anamnia is most active dorsally, giving rise to the bulky dorsal region of the body.

The Lower Chordates.

Among the lower chordates the question of growth in length has been considered with respect to Amphioxus by a number of investigators. Hatschek originally described growth in length as being due chiefly to the activity of two large cells which he termed polar cells, and from which he supposed all the hinder mesodermal segments were formed. Others, Wilson, Morgan, etc., have shown that these polar cells were mythical and they described growth in length as due to a general mass of proliferating tissue as in the higher chordates.

Lwoff, Cerfontaine and others have also worked at the subject.

Hatschek and Cerfontaine have maintained that growth in length is due to concrescence of the dorsal lips of the blastopore in precisely the way that we have shown to be quite untenable for the Teleostean and Elasmobranch fishes.

Lwoff described the dorsal wall of the hinder part of the embryo as forming from the dorsal lip by a process of inrolling of the ectoderm, this giving rise to what he called an ecto-blastogenetic plate.

MacBride has more recently gone into the matter and has come to the conclusion that there is no such thing as a concrescence of the dorsal lips of the blastopore, but that increase in length is due to a growing point exactly comparable to that which I have already described as a deuterogenetic centre, due to the general activity of the lips of the blastopore. MacBride however considers that the closing of the blastopore is very largely due to the growth upwards of the ventral lip of the blastopore; a condition which does not fall very well into line with the method by which the Frog's blastopore closes. Here, as in the anamniate craniates, the more active part of the blastopore rim is the dorsal.

This fact according to MacBride becomes evident very early, even while gastrulation is in progress (Fig. 8).

In the same way as in the higher chordates the ventral part

of the deuterogenetic centre dies out after a time, and the dorsal and dorso-lateral part continues to grow, giving rise to the tail.

In the Ascidians the conditions are much the same, as shown by Conklin in the development of Cynthia. Here the activity of the growth of the ventral lip is perhaps more marked than in Amphioxus.

In the Hemichordata the question has not been considered very thoroughly.

It is well known that no part of their body can be compared with the tail of the higher chordates.

If we are to compare their condition with that of the higher chordates, we must suppose either that

(i) The deuterogenetic centre dies out wholly almost as soon as formed, and that all growth is interstitial and protogenetic; or,

(ii) The deuterogenetic centre continues active all round the blastopore, which however closes, with the result that the anus is at the extreme end of the body, and that there is no tail.

The egg of the Balanoglossidae forms a hollow blastula, which invaginates as in Amphioxus and forms a gastrula. The gastrula elongates. The blastopore closes but the anus forms at the spot where the blastopore has closed. An anterior unpaired pouch and two pairs of lateral pouches are given off from the gut which form mesodermic sacs, the five cavities which are known as the proboscis or head cavity, the collar cavities and the trunk cavities. There is no segmentation of the trunk cavities as in Amphioxus although the trunk grows out to a far greater extent.

This latter is an important point for it is not unreasonable to suppose that segmentation may be due to growth being from a definite terminal growing point.

If growth in length is produced like the tail of a vertebrate is produced, or like a stem of a tree is produced, it is clear that although median organs may be indefinitely prolonged by this means, and yet retain their original characters, lateral organs cannot be indefinitely prolonged and at the same time continue to act as lateral organs.

A notochord, or a spinal cord, can be produced indefinitely, and these organs do not show segmentation.

But a lateral blood-vessel, or a lateral nerve, or a branch of a tree or a leaf could not be produced in the same way and yet retain their original character. They can however be repeated—and indeed we find in organisms in which growth in length occurs in this way that lateral organs *are* repeated, and not drawn out —and this fact may be at the bottom of metameric segmentation.

If so it follows that metameric segmentation may have arisen independently in many different phyla and that it is a result of growth in length from a definite growing point; and as the growing points may have had different origins, there need be no homology between the metamerism of one phylum, for instance the Annelida, with that of another phylum, for instance the Vertebrata.

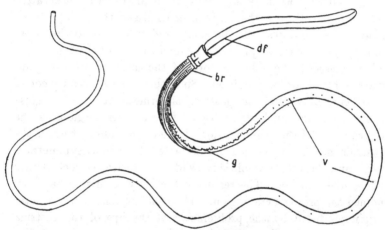

Fig. 30. *Dolichoglossus serpentinus.*
df dorsal furrow of proboscis; *br* gill clefts; *g* gonads; *v* bright vermilion spots.

As regards Balanoglossus it seems probable that there is no deuterogenesis, or if it occurs it soon dies out altogether. The great increase in length of the trunk is due to general interstitial growth. This seems probable from the fact that the bright spots which occur on the trunk of Dolichoglossus become gradually more and more separated from one another as the organism elongates, Fig. 30.

The bright vermilion spots are the remnants of pigment formed in the gonads, and probably of the nature of excretory products. The gonads, however, are closely packed against one another, whereas the spots towards the end of the trunk are perhaps an inch or more apart. This seems to prove that their separation and hence the growth in length is due to a general interstitial growth and not to a terminal area of cell proliferation.

If deuterogenesis occurs it occurs all round the anus (or blastopore) and there is no tail. The absence of tail is a very important fact, for it is in the Hemichordata alone among chordates that there is no tail.

This is so important that it would justify the classification of the Chordata into *Chordata caudata,* and *Chordata ecaudata.*

Conclusion.

To sum up we may say, both from anatomical observation and from experiment, that growth in length of the embryo in all anamniate chordates must be considered to be due to the origin of a special area of cell production round the lips of the blastopore, which converts the spherical form of the gastrula into the cylindrical form of the later embryo. Since this area of necessity comes into being only after the gastrula is formed, we may recognise two regions in the later embryo. One of these regions is the direct result of the segmentation of the ovum culminating in the gastrula, and having the general character of a radially symmetrical form, and this on the whole is to be identified with the coelenterate phase of evolution. The region of the body so arising has been named the protogenetic region. The other region is that of later origin produced by the proliferation of the lips of the gastrula mouth. This has been called the deuterogenetic region. The part formed from the protogenetic region includes the forebrain, probably also the mid-brain, the mouth, and possibly the hind-brain as far as the origins of the 5th and 8th nerves, the branchial region and heart and probably much of the gut. The part formed from the deuterogenetic region comprises the remainder of the hind-brain and spinal cord and tail, the whole of the metamerically segmented mesoderm, and *in the craniates* the renal organs. As regards the reproductive organs there is much evidence to show that in the craniate chordates the actual germ cells are, as one would expect, protogenetic in origin, but that they migrate during development into the deuterogenetic region and here undergo their maturation, and eventually find their way to the exterior by means of the deuterogenetic channels of the coelom or renal apparatus.

The same relations between the two regions probably hold good for the amniotes, though in them experimental evidence is obtained less easily.

Now it is conceivable that having attained the stage of a coelenterate the gastrula may take very different courses of development. For instance, the whole deuterogenetic centre might continue active, and thus produce a long cylindrical addition to the spherical gastrula, Fig. 31 *A*. Or it might cease almost

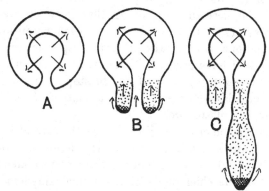

Fig. 31. Diagrams of sagittal sections of stages in the development of a craniate chordate, showing the growth centres.

A, Gastræa stage; *B*, Balanoglossus stage; *C*, tailed chordate stage.

as soon as formed, in which case there would be no great alteration from the radially symmetrical form, Fig. 31 *B*. The organism might retain more or less perfectly its radially symmetrical form. This would be due to the dying out altogether and at once of the deuterogenetic centre of activity. Or a portion of the deutero-genetic region might remain for a longer time active while the rest dies out earlier, Fig. 31 *C*. The portion that remains active might be that forming one border of the blastopore, while all round the rest of the blastopore the activity dies out. The latter part therefore would be left stationary while the active part grew back.

It may be suggested that deuterogenesis in the embryo is a geometrical consequence following gastrulation (which may

perhaps also be regarded as a necessary culmination of protogenesis). If so these consequences of embryonic development
must have had their effect upon evolution. For instance, one
of the first effects of deuterogenesis must be in many cases a
tendency to close the blastopore.

Now clearly in evolution to close the only entrance to the
digestive tract would be fatal. Therefore there must always have
been and must always be the counteracting circumstances of the
"necessity to live" opposing the geometrical tendency, just as
there always are and always have been the counteracting
influences of the force of gravity which would bring us to the
earth, and the necessities of living that keep us erect. The
continual warring of these tendencies at each repetition of the
life cycle, the one to close, the other to keep open the gastrula
mouth, may well have given the impetus in evolution which has
led to the very varying fates of the blastopore in the different
groups of animals, and the consequent relation of main axis to
the plane of the blastopore.

The two phenomena, the difference in the fate of the blastopore, and the difference in the fate of the deuterogenetic centre,
are of fundamental importance in our attempt to understand the
relationship of the chief phyla of the animal kingdom. Let us
take a very brief survey of what we know of the actual fate of
the blastopore in individual species. In *Coelenterates* the fate is
simple. The blastopore remains open, acting as chief inhalent
and exhalent aperture. The deuterogenetic centre, if there ever
is one, dies out wholly and at once. In *Echinodermata* the blastopore may be said to remain, but as anus only. A new opening
is formed, which is the mouth. The deuterogenetic centre probably
dies out wholly and at once. In the *Chordata* the blastopore remains in part as anus or else the anus opens where the blastopore
has closed. We may regard this as a re-opening of a part—the
ventral part—of the blastopore. To be accurate it is the more
ventral part of the blastopore which is anus, the dorsal part, after
becoming enclosed by the neural folds forms the neurenteric canal,
which ultimately closes. The mouth is never formed from any
part of the blastopore; it is always a new opening. The whole
deuterogenetic area continues active for a short time, then the

dorsal portion alone continues, with the result that a tail is formed. This I have endeavoured to show is the only conclusion to which we can come as the result of experimental and anatomical observation in the chordate embryos. Therefore the relation of the main axis of the Vertebrata is, as I have indicated, at right angles to the plane of the blastopore in Fig. 31 *C*.

This conclusion is diametrically opposed to the conclusion of His, Hubrecht, Dohrn, Semper, Sedgwick and others, who derive the dorsal surface of the chordate from the coalescence of an elongated blastopore surface.

So even MacBride, who after fully accepting as true the views of growth in length as I have indicated them, comes to the conclusion, that "the mouth and anus in all animals in which two such openings are found, owe their origin to the division of a long slit-like coelenterate mouth," and suggests that the seam which originally connected mouth and anus was situated "on the *ventral* and not on the dorsal surface."

This supposition has not a trace of embryological evidence in its favour.

Goodale's experiments already alluded to show that there is no suggestion of concrescence either along the ventral or dorsal border connecting mouth and anus.

MacBride suggests that the growth upwards of the ventral lip of the blastopore in Amphioxus is reminiscent of this concrescence. It is true that a similar and more obvious coalescence of the ventral part of the blastopore lip occurs in the Amphibia, but the most ventral part either remains as anus or re-opens as anus— never is there any sign of connection with the mouth. We might legitimately regard the Amphibian condition as representing an ancestral coelenterate or actinian phase in which the neurenteric end or opening represents the actinian inhalent channel; whilst the anal end or opening represents the exhalent channel.

One may argue thus. When the chordate central nervous system began to become bulky and in forming a tube involved the inhalent channel (as neurenteric channel), then one or more of many other openings which in no way owed their origin to the blastopore became an inhalent aperture or mouth. There is no difficulty in assuming the adoption of a new opening for the mouth either in the Chordate phylum or in the Echinoderm phylum. In both these phyla openings from the gut or its pouches are common—as they are in the Coelenterate. In the coelenterate we find numerous apertures in the tentacles and in the body wall, as for instance the cinclides of Anemones. In Chordates apertures of this nature are numerous, e.g. the gill clefts of any class; or the intestinal pores of certain Hemichordata.

In the Echinoderm also the mouth is the new opening. The hydropore is also a new opening into an archenteric diverticulum. It may be objected that gill clefts are essentially exhalent and not inhalent apertures. No one, however, would deny the homology

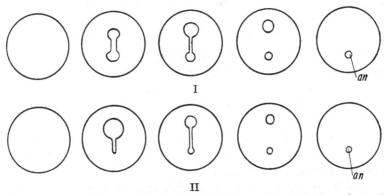

Fig. 32. Diagram to show the closure of the blastopore, and to show the part that remains open as anus, *an.* I, Urodela; II, Anura.

of the spiracles of the Dog-fish with the branchial clefts. It is almost certain that they must at one time have been exhalent apertures. Now they are inhalent. There is no reason why the present mouth should not have arisen from either an inhalent or exhalent pore, provided it occurred in a suitable place.

In both the Chordata and Echinodermata we find strong embryological evidence for the assumption that the blastopore became anus only and that the present mouth is a new pore. In every case described the anus is always either derived directly from the blastopore or else it opens at the spot where the blastopore has closed. Thus the Chordata and Echinodermata agree in their main axial relation to the plane of the blastopore, for, in each case it is at right angles to this plane. They also agree in various other respects. But the two phyla differ in respect to deuterogenesis. In the Echinoderms it died out—such elongation as there is is probably interstitial, as possibly in Balanoglossus. In the caudate Chordates it continued active dorsally, and as already described produced the tail, see Fig. 31.

Now it has probably been very different with regard to the other phyla. Unfortunately at present there is no experimental evidence to help us, but a good deal of accurate anatomical evidence is available.

What is the fate of the blastopore in Annelida? The evidence suggests that there is an elongation of the original gastrula mouth in the plane of the blastopore, which elongated mouth becomes reduced by a coalescence of the middle lips into two terminal openings (as in the Urodele Amphibians) forming mouth and anus respectively.

And further it would appear that the deuterogenetic centre remains active between the two openings, but more particularly near the anal opening producing growth in length. The direction of which elongation is therefore at right angles to the direction of the Chordate elongation.

Thus the fate of the blastopore in *Hydroides uncinatus* (Eupomatus) is described thus by Shearer: "The blastopore at first lies exactly in the middle of the ventral plate...when fully formed it is an elongated slit, somewhat enlarged at its anterior end. This end never completely closes, but...becomes the mouth. The posterior portion closes completely, the anus breaking through almost immediately where the last portion of this part of the blastopore disappears." So also in Polygordius. Conn says of serpula: "the blastopore becomes an elongated slit, the lips of which meet in the middle and close, forming the rudiment of the future gut. For a short time the digestive tract remains attached to the ectoderm throughout the length of the blastopore, but after a little it only retains this connection at either end. The digestive tract becomes hollow and acquires two openings to the exterior at the two points of its previous connection with the ectoderm."

In *Amphitrite ornata* according to Mead the blastopore closes by concrescence from behind forwards but the extreme anterior end remains open as the mouth, and the anus breaks through at the posterior end of the seam.

Balfour says that Stossich and also Willemoes-Suhm state that in one species of Serpula the part of the blastopore remaining open is the anus.

In Oligochaeta the blastopore becomes the mouth.

This can all be easily rendered intelligible by assuming that in evolution the blastopore elongated and then by closing centrally gave rise to the mouth and anus.

Growth in length in this case has been due to the persistence of part of the blastopore rim as the deuterogenetic centre. But which part? Clearly not the mid-dorsal as in Chordates. It is probably caused by the retention of a part of the coalesced lip. This is the part which has been sometimes named the teloblast. It gives rise to the trunk and the so-called coelomesoblast; and as it is situated between the mouth and anus the effect of its activity is of course to separate these two openings more and more.

In *Hydroides uncinatus* (Eupomatus), Fig. 33, the cells which give rise to the coelomesoblast from which the greater part of the trunk mesoblast is derived have their origin in the lips of the blastopore, as Shearer has shown, or as Treadwell has shown in Podarke, near to the future posterior end.

In many Annelids (and Molluscs) there are other mesodermal cells which are not derived from these two blastopore lip cells, but either from the ectoderm or other part of the endoderm. These form the so-called larval mesoblast or ectomesoblast.

If we use the same terminology as that which we have used with regard to the Chordates, then clearly the ectomesoblast is protogenetic mesoderm, and the coelomesoblast in these forms is deuterogenetic. The growth in length is due to deuterogenesis but it is due to the persistence of a different part of the deuterogenetic ring, and the result is a lengthening in the plane of the blastopore in Annelids, and not at right angles to it as in Chordates.

Moreover in Annelids both mouth and anus have had their origin in the gastrula coelenterate mouth, whereas in Chordates only the anus has been derived therefrom, Fig. 31.

The group most closely allied to the Annelida is undoubtedly the Arthropoda. In these there is plenty of indirect evidence that the plane of the gastrula mouth is the plane of the future ventral surface. For instance in the Copepod Cetochilus the blastopore is a longitudinal slit on the future ventral surface and it closes from before backwards, the part which closes last being

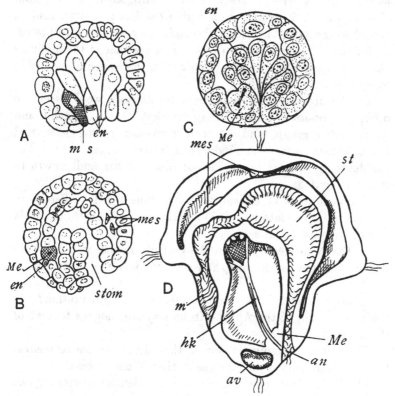

Fig. 33. *A, B,* Podarke, after Treadwell; *C, D, Hydroides uncinatus,* after Shearer.

Semi-diagrammatic views to show the derivation of mouth and anus openings. *an* anus; *mes* ectomesoblast; *en* endoderm; *Me* coelomesoblast; *stom* stomodaeum; *m* mouth; *hk* head kidney; *st* stomach; *av* anal vesicle.

near the site of the future proctodaeal opening. In Limulus and the scorpion or in insects such as Doryphora there is a median ventral furrow which represents the blastopore; and above all there is the evidence of the intermediate form Peripatus, where the blastopore actually elongates into a long narrow slit which closes in the middle region but the ends remain permanently open as mouth and anus.

So in the Mollusca in some, as for instance, Ostrea, the blastopore remains open as the mouth, in others, Pisidium, it remains as the anus, in others, Teredo, it closes altogether. Such inconsistencies taken in conjunction with what has just been said in regard to the Annelida and Arthropoda may well be interpreted, as Balfour suggested long ago, as indicating that in the Molluscs also the gastrula mouth has given rise in the course of evolution to both mouth and anus.

In the Mollusca the deuterogenetic activity would seem to occupy a position corresponding to that of the Annelid and Arthropodan group, and we can recognise protogenetic and deuterogenetic mesoderm produced in a very similar manner, but the deuterogenetic activity dies out much earlier and growth in length is limited.

The conditions of development are thus totally different to those of the Chordata and Echinodermata in which the gastrula mouth gave rise to anus only.

In the Nemertean phylum the gastrula mouth would appear to have become mouth only, and the anus is a new formation.

It may be asked what about metameric segmentation? Is the metamerism of the Annelid in no way homologous to that of the Chordate?

To this I would reply, no more so than the metameric segmentation of the Annelid is homologous to that of the fir tree.

The similarity is one of analogy. The deuterogenetic regions of the Annelid and of the Chordate show metameric segmentation because in each case growth in length is due to the same physiological cause, which is quite analogous to that which produces growth of the stem of a fir tree. Axial organs can be indefinitely produced in length without interfering necessarily with their utility. Thus the intestine of an earth-worm, the notochord or spinal cord of a vertebrate, or the pith of a fir tree, can be indefinitely extended and yet retain their original functions.

But we cannot imagine lateral organs like lateral blood vessels, lateral nerves, lateral muscles, limbs, leaves on a stem, being expanded indefinitely without altogether losing their utility, so such organs must be repeated and not expanded when the great growth in length occurs. Hence it is not necessary to regard metamerism as in itself indicating homology.

The object of this short course of lectures has been to show what the real relation of the main axis of the body of a Chordate is to the plane of the gastrula or coelenterate mouth, and to indicate that it is probably quite different to that of the Annelid, Arthropodan and Molluscan groups. Such a conclusion if correct is quite destructive to the hypothesis of His, Semper, Dohrn, Patten, Gaskell and others who would derive chordates from various highly organised invertebrate groups. It agrees with the views of Sedgwick, van Beneden and Hubrecht in seeking the origin of chordates as far back as the Coelenterata or their immediate descendants, but is equally destructive of the view that the main axis of the chordate body is parallel with the plane of the coelenterate mouth—on the contrary it shows it to be at right angles thereto.

THE GEOMETRICAL RELATION OF THE NUCLEI IN AN INVAGINATING GASTRULA (e.g. AMPHIOXUS) CONSIDERED IN CONNECTION WITH CELL RHYTHM, AND DRIESCH'S CONCEPTION OF ENTELECHY

PART I

"No kind of causality based upon the constellations of single physical and chemical acts can account for organic individual development....Life, at least morphogenesis, is not a specialised arrangement of inorganic events; biology, therefore, is not applied physics and chemistry: life is something apart, and biology is an independent science."

"The Science and Philosophy of the Organism" by Hans Driesch, being the Gifford Lectures for 1907, p. 142.

Although one may subscribe to the opinion expressed in the above quotation from Driesch, nevertheless on reading the whole book it strikes one that Driesch's "Entelechy" is something much more than the minimum contained in that statement, and is indeed a somewhat mystical conception.

To fill in the sentences omitted in the above quotation of Driesch, the omission of which was indicated by the asterisks: "This development is not to be explained by any hypothesis about configuration of physical and chemical agents. Therefore there must be something else which is to be regarded as the sufficient reason of individual form-production. We now have got the answer to our question, what our constant E consists in. It is not the resulting action of a constellation. It is not only a short expression for a more complicated state of affairs, it expresses a *true element of nature.*"

Then again later "our vitalistic or autonomous factor E concerned in morphogenesis" is named "Entelechy," p. 144. But Driesch's Entelechy does seem to me to be a more complex conception than that of a "true element of nature" or a form of energy in any way comparable to the single physical forces. It is a

5—2

coordinating and regulating influence with very diverse and mysterious ways and means of application, not unlike the recapitulatory "constraint" which twenty years ago was held to *cause* organisms to develop along a certain ancestral course, and is amenable with difficulty, if at all, to exact computation.

While welcoming the conception because it involves the postulation of some form of energy not present in inanimate bodies, one may ask whether for the present it would not be more profitable to go no further in the search for the unknown vitalistic factor than the recognition and investigation of a special form of energy not so wholly different from certain known forms of energy.

Such a form of energy might be conceived of as acting from a centre like gravitation, statical electricity or magnetism; as causing movements of attraction or repulsion according to the conditions under which it is acting; as being a constant attribute of living matter and exerting no influence on non-living material; and as possessing the peculiarity of automatic and rhythmic alternation between unipolar and bipolar states.

May not Driesch's "Entelechy" prove to be the more complex action of some such simpler form of energy in combination with other forces?

May not Entelechy bear to this simpler form of energy some such relation as that, for example, which the gyroscope bears to gravity, and the real vital factor be something rather less complex and also easier to investigate by exact methods of mensuration and computation?

In 1894 Roux made the discovery that the blastomeres of cells of *Rana fusca* towards the end of segmentation, if isolated and floated in a suitable medium, will show distinctly attractive properties *inter se*, so that the sides of each cell will become drawn out towards a closely neighbouring cell, and such cells, eventually moving towards each other and coming actually into contact become pressed and flattened up against one another. This phenomenon Roux termed cytotropism, which he has since. changed to cytotaxis, a term also adopted more recently by Przibram.

If this alleged attraction is a reality—and if it is a universal law, then the discovery of this fact must be regarded as one of great importance and of far-reaching consequence. Indeed this property has already been recognised and called upon as an explanation of certain developmental phenomena as for instance by Hjalmar Théel who observed that apparent attractions and repulsions occur among the blastomeres of the segmenting egg, *Echinocyamus pusillus*.

Zur Strassen especially, has endeavoured to apply the principle of chemotaxis and cytotaxis to the formation of simple epithelia in general and the blastula epithelium of the developing egg of *Ascaris* in particular.

He suggests that a simple, single-layered epithelium of cubical or columnar cells, supported by a sub-lying tissue or membrane, may be the result of a chemotactic attraction between the basement membrane and the several cells producing a result such as would be produced by placing a pile of shot on a table and then shaking until the shot all settled down into a uniform unilaminar layer.

For the production of more complicated cases he supposes a much more complicated series of attraction zones, so that one part of a cell attracts its neighbour more strongly than the next part—the parts beyond again still more strongly and so on, and thus he accounts for the curves seen in various "free" epithelia.

When some years ago I originally came to believe in the phenomenon now known as cytotropism or cytotaxis (though perhaps cytokinesis would be a still more appropriate term), I arrived at it as the result of experiments made with a view to determine the cause of invagination such as occurs during the process of gastrulation in *Amphioxus*.

The mechanics of the invagination process as seen in *Amphioxus* during gastrulation have been often discussed, e.g. Hatschek, Goette, zur Strassen, Rhumbler, etc.

Rhumbler in his recent analysis of the factors involved in the process of gastrulation by invagination distinguishes between the factors which increase invagination when once initiated and those which are the actual cause of it. In his model made of flexible steel rings he perceived that there would have to be an actual force which acts as for instance one's finger acts in producing an invagination in an indiarubber ball, either a pushing from without or a pulling from within which initiates the invagination. Once started, then lateral pressure caused by the multiplication of cell units will increase it; but without forces other than those which he supposes to exist, lateral pressure cannot initiate invagination.

It has been suggested that invagination of a gastrula such as that of *Amphioxus* may be brought about by the absorption of the fluid within the blastocoel, causing the collapse of one side. Hatschek suggested this as a possible cause in *Amphioxus*; but apart from the unlikelihood of the thicker side being the invaginated side as is actually the case in *Amphioxus*, if this were the cause, it has been completely put out of question by the fact noted by Morgan and Hazen that invagination will occur when a small aperture exists in the blastula wall.

Rhumbler like Goette points out that the shape assumed by cells forming the wall of a blastula is conical, with the smaller end pointing inwards, but that the cells of the invaginated side of a gastrula are conical with the smaller ends pointing towards the original outside, and Rhumbler suggests that a change in shape of the cell has been the cause of the invagination. Surely this is confusion of effect and cause. The living cell in an undifferentiated state is an almost perfectly fluid body and the rectangular or hexagonal shape assumed in early undifferentiated tissues is much more easily to be explained as the result of environmental pressure than as being due to its own intrinsic properties. And does not a cell isolated from blastula or gastrula tend to assume a spherical form?

He argues, however, that the change in shape is due to the absorption of fluid by the inner portions of the cells causing thereby a bulging of their inner surfaces. This, occurring in many or most of the cells of one part, would, he says, produce the change in shape, and together with the lateral pressure due to multiplication of the ectoderm units would bring about invagination.

Przibram commenting on this and other suggestions offered concludes "Blastulation and gastrulation depend on chemotactic effects started by processes of assimilation, which not only cause passive mechanical displacements, but also active migrations of cells."

But beside the objection raised above that the conical shape of the cells would seem to be more the effect than the cause, and the subsequent change of shape during invagination likewise an effect rather than a cause, it is very doubtful whether, if the changes in assimilation and surface tensions suggested do occur, they could produce a swelling of the inner part of the cell and corresponding reduction of the outer, because we know that the cell when isolated tends to assume a spherical form; that is to say cells do not retain a conical shape after the mutual pressure has been removed, thus showing that they have no intrinsic tendency to be conical (v. Herbst's figures), and so can hardly be so prone to become conical in the obverse direction as to be able to exert the pressure necessary to convert a blastula into a gastrula.

In my original experiments on the process of gastrulation in *Amphioxus*, I experimented with indiarubber balls, which were supposed to represent isolated cells, and which like cells, when pressed together become flattened. In most respects an india-rubber ball behaves as an isolated blastomere does when at rest. I tried by various means representing increase in number illus-trating, that is to say, more rapid growth at various places, or by variation in the size or shape of ball, to reproduce gastrulation such as occurs in a typical way in *Amphioxus*, but by no means could I do this until I used indiarubber cord applied so as to represent a mutual attraction between cell and cell.

By means of a circle of indiarubber balls strung together by indiarubber cord I could reproduce exactly the process of con-version of a blastula into a gastrula, or to be more accurate, the conversion of a circle into a double crescent, one part of the circle becoming "invaginated" into the other.

The necessary conditions are: (i) that the balls shall be sufficiently numerous to cause tension in the elastic cord—the balls themselves then becoming flattened against each other, the free inner and outer surfaces remaining convex.

(ii) The cord must pass nearer to the outward surfaces of a certain number of the balls along one part of the circle than elsewhere. That is to say, if the cord pass through the centre of the balls for two-thirds of the circle, and pass, say half way between the centre and the outer surfaces of the balls forming the remaining third, then that remaining third will be invaginated, if the third and last condition is fulfilled, namely,

(iii)   That the whole circle is sufficiently large; because the invagination process can be inhibited if the circle is so small that the pressure between the neighbouring balls at the point of contact

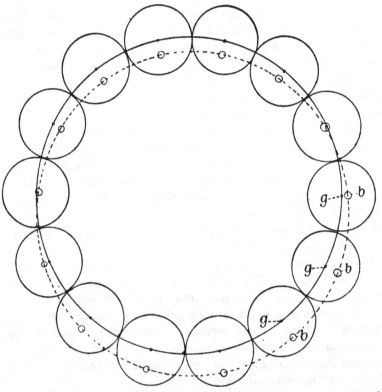

Fig. 34.   Diagram representing a circle of spheres in which the relative position of the centres of attractions (*b*) to the centres of the mass (*g*) necessary to cause invagination of one part of the circle within the other part are shown.

is greater than the attractive force between the attractive centres of ball and ball, (which *ex hypothesi* are eccentric to the whole mass of the ball over a certain small area of the sphere or arc of the circle,) in this case no invagination will take place.   But increase the number of balls and then the invagination may occur. The accompanying diagrams show how this must be so.

Fig. 34 represents a series of spheres arranged in a circle. Each sphere has its actual centre indicated by the black spots *g, g....* And each sphere is supposed to be made partly at any rate of living material and to have a centre from which it exerts an attractive force upon its neighbours. This centre, *b, b...* is shown not to correspond with the actual centre, but to lie near the internal pole of the spheres forming the upper part of the diagram, and nearer to the outer margin in the spheres forming the lower part of the diagram.

Under these circumstances the form taken by the ring of spheres would not be a circle, but one side—the lower side would be invaginated within the upper—provided that the attractive force between cell and cell (contiguous cells) is strong enough.

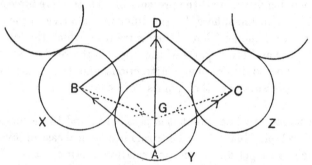

Fig. 35. Parallelogram of forces concerned in the invagination of a circle of spheres such as shown in Fig. 34.

Fig. 35 illustrates in detail the action of the forces exerted between cell and cell of the lower part of the diagram.

Instead of taking the whole lower part into consideration, I have taken only three spheres, in the middle one (*Y*) of which the centre of attraction (*A*) is shown to be near to the outer margin of the circle while in the two side ones (*X* and *Z*) it is co-incident with the centre of the spheres at *B* and *C*.

The parallelogram *ACDB* will be the parallelogram of forces exerted upon *A* by the attractive powers of the spheres *X* and *Y*, and *Z* and *Y*. The lines *AB, AC,* may be taken to represent the direction of these forces and the line *AD* the resultant.

If the tissues are perfectly compressible the sphere *Y* must move in the direction of *D*.

But if the tissues are only imperfectly compressible there will then be a counteracting force exerted on Y by its neighbours at the points of contact, tending to prevent the movement of Y towards D. The strength of this counteracting force will depend *inter alia* on the position of these points of contact. The nearer they are to the point D, the greater will be their influence, the further removed from D, the less will be their influence and the less will be the tendency to prevent the movement of Y towards B. Now clearly with a sphere (or ring of balls as in the diagram) any enlargement of the sphere (or ring) will move the points of contact further from D and diminish the counteracting force due to physical pressure and will therefore favour the movement of Y towards D. Thus the invagination of the lower pole of the diagram depends upon (1) the presence of an attractive force acting between sphere and sphere, (2) the difference, relative to the centre of the mass, of the position of the centre from which the force acts in the upper and lower parts of the ring, (3) the relative strengths of the supposed inter-spheric attractions and the counteracting physical pressure at the points of contact of sphere and sphere.

Let us see to what extent it is possible to find these conditions in the developing blastula and gastrula in a typical case of invagination such as we get during the gastrulation of *Amphioxus*. In the first place we must regard the phenomenon of intercellular attraction as an established fact, and believe that an attraction exists between cell and cell as an attribute of a living cell, as much as gravity is an attribute of all matter.

Like gravity we must consider that it can be described as acting from a centre. Just as there is a centre of gravity for every mass, and just as the centre of gravity may be not the actual centre of the mass, so there must always be an attraction centre in a cell, and this attraction centre need not correspond to the actual centre of the cell mass, nor to its centre of gravity. Thus in a bird's ovum for instance, which we may consider to be a sphere, the centre of gravity is certainly below the actual centre of the sphere (i.e., nearer the "vegetative" pole) while its attraction centre will probably be very much above the actual centre of the sphere.

Given these conditions, then the mechanics of the rubber ball model may be reproduced by the living blastula and gastrula.

The next question to be considered is, can we locate the attraction centre?

There can be little doubt that if it exists, it is coincident with either the nucleus (Hertwig's law "The nucleus tends to take up a position in the centre of its sphere of influence, i.e., of the protoplasmic mass in which it lies," Przibram), or more probably still, the centrosome of the resting cell. The centrosome is, however, not always easy to see and at any rate it is not often drawn in figures of sections of developing embryos, but the nucleus is usually given, and as the nucleus is never far removed from the centrosome we may for the purpose of the argument take the centre of the nucleus as indicating approximately or perhaps actually the position of the centre of the attractive force. When I first endeavoured to work this out, the only figures available were those of Kowalewski and Hatschek, and those of Hatschek were regarded as authoritative. Now unfortunately for my hypothesis the position shown by Hatschek for the nuclei did not support in any way the contention that the supposed attraction centre of the invaginating cells was more towards the outer surface than the inner surface as was necessary to fulfil the required conditions. On the contrary just the opposite was shown, e.g., v. Hatschek, Fig. 21, though not in all cases, e.g., Fig. 20, 2 b.

But although Hatschek's figures are unsurpassed for clearness and general accuracy, yet as regards the position of nuclei, and of cell walls and in some other respects, they must be regarded as diagrammatic, as recent work on *Amphioxus* has shown.

During the last fifteen years sections of the gastrulating *Amphioxus* have been drawn by several authors which are probably more accurate in this respect, because they agree with one another, though disagreeing with Hatschek. They all agree in showing the nuclei of the cells about to invaginate and the invaginating cells as lying close up to the outer surface of the cells, and what is more to the point in most cases distinctly more eccentric than the nuclei of the cells which do not participate in the actual invagination. I allude to the drawings of Sobotta, Samassa, Morgan and Hazen, Wilson, MacBride, Cerfontaine and Legros.

I reproduce here figures from Sobotta, Samassa and Cerfontaine which illustrate my point as well as possible (v. Fig. 36).

In Samassa's figure *A*, for instance, the difference of the position of the nuclei as regards its eccentricity is represented by the figures 1 : 1·73 for the fourteen uppermost cells compared with 1 : 2·40 for the twelve lower larger cells, the figure 1 representing the average distances of the centre of the nucleus from the outer margin of the cell, the figures 1·73 and 2·40 the average

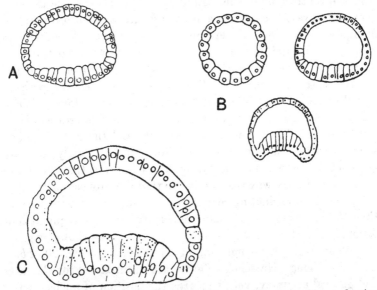

Fig. 36. Drawings of sections of the gastrulating blastula of *Amphioxus*, showing how the nuclei of the cells occupy the position within their cells required by the hypothesis explained in the text. *A* after Samassa, *B* after Sobotta, *C* after Cerfontaine.

distance of the centre of the nucleus from the inner margin of the uppermost 14 and lowermost 12 respectively.

It is a very remarkable fact that in all these figures the nuclei of the invaginating cells occupy the position required by the hypothesis. Moreover a similar state occurs in many other gastrulating blastula, as, for instance, may be seen in the figures of *Synapta* by Selenka, *Balanoglossus* by Bateson, etc.

The condition that determines the position of the attraction centre is probably the degree of concentration of the protoplasm.

If there is a form of energy which is an attribute of living cells and which under certain conditions is manifest as an attractive force acting from a centre, the "centre" must be the centre of the living substance, and if the protoplasm is vacuolated through part of its mass by dead material, for instance, food yolk, then the centre will be driven away from the yolk mass just as the centre of gravity is driven away from the centre of a mass by vacuolation such as one may get in part of a lump of slag. Further it must be supposed that the divisions of the blastomeres take place in such a way that the more yolk-laden parts of the cells are directed

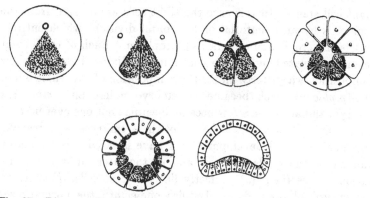

Fig. 37. Diagrams illustrating how a mass of inert material within an ovum causing eccentricity of the centre of attraction may be divided in the vertical plane without disturbing the original proportion of its distribution between upper and lower segments of the segmented ovum, but how a horizontal plane of division upsets this arrangement.

inwards, and that therefore the disproportion between the segments set up by the first horizontal furrow, that is to say, at the third generation of blastomeres, is retained till the end of segmentation.

The material which vacuolates the protoplasm and causes the eccentricity of the centre of attraction in the case of *Amphioxus* is no doubt the food yolk, small though the amount of it may be.

It is clear that on the above hypothesis the distribution and quantity of yolk relative to the upper and lower poles, is an all important factor in the process of invagination of the blastula, and anything which would interfere with the normal distribution of yolk would probably render the invagination imperfect or impossible.

This is very prettily shown by Wilson's experiments on the blastomeres of *Amphioxus*. Wilson found that by shaking apart the blastomeres formed by the first cleavage furrow, each separated blastomere developed normally and became a perfect blastula, gastrula, and metameric embryo, but half the size of a normal one. So also each of the first four segments became perfect gastrulae and free swimming embryos. Now in both these cases the plane of division is vertical and median and divides the egg in such a way that the relative distribution of the protoplasm and yolk between upper and lower poles is undisturbed.

But a very different result was obtained by the segments of the eight cell stage. In this case the eight cells are derived from the four cells by an equatorial or horizontal division, which entirely upsets the proportion of yolk to protoplasm. Each of these eight segments isolated by Wilson continued to divide, but since their composition was utterly unlike the composition of the whole ovum of *Amphioxus*, each became an embryo unlike an *Amphioxus* embryo, and although a few became blastulae not one ever became a gastrula. Wilson says: "None of the $\frac{1}{8}$ embryos, as I believe, are capable of full development. I have isolated a considerable number of the $\frac{1}{8}$ blastomeres, and of the later embryos of the various types (i.e., '(*a*) perfectly flat plates of cells, (*b*) more or less curved plates and (*c*) blastulas one-eighth the normal size, either closed or with a pore at one side'), and have observed hundreds of all stages without once obtaining a gastrula."

Morgan a few years later repeated these experiments and he gives on the whole a similar, but not so dogmatic an account as Wilson's. Thus he says some of the $\frac{1}{4}$ blastomeres fail to gastrulate while a few of the $\frac{1}{8}$ blastomeres form blastulae which "partially gastrulate." "Many of the $\frac{1}{8}$ blastomeres develop into hollow, swimming blastulae, which swim around for several hours after the normal blastula has gastrulated. Some of these blastulae flatten at one pole as though about to gastrulate, but do not seem to be able to develop further."

The shaking of the eggs however seems to cause abnormalities even in the whole ova.

I think we may safely conclude that whereas the rule is that $\frac{1}{2}$ and $\frac{1}{4}$ blastomeres develop as the whole ovum normally does, the $\frac{1}{8}$ blastomeres will not invaginate. Again, Wilson did find some very small gastrulae "even less than $\frac{1}{8}$ the normal size," but he says "these gastrulas however did not arise from 8-celled stages but from 2- and 4-celled, and I can only explain their origin by supposing either that the $\frac{1}{4}$ or $\frac{1}{2}$ blastomeres underwent preliminary fission before beginning their progressive development, or that they were mechanically broken into smaller fragments by the operation of shaking as often happens in the case of entire undivided ova."

Thus a gastrula $\frac{1}{8}$ or $\frac{1}{16}$ normal size may be obtained, but these are derived from $\frac{1}{2}$ half embryos or $\frac{1}{4}$ half embryos, in all of which cases the reduction is *due to vertical partitions* which do not disturb the proportions of upper and lower poles as a horizontal partition does. On the hypothesis which I suggest these observations receive a perfectly comprehensible explanation.

Granted that there is such a form of energy we must inquire whether the attractive property is constant or whether it is only temporary or whether it is varying in intensity. For instance, it might be more intense after the complete formation of each cell or blastomere, and gradually diminish with the age of the cell, while passing gradually into the bipolar state to become again more evident as an attractive force after the recurrence of cell division.

Anyone who is familiar with the fresh segmenting ova of mammals will remember how when first found—presumably while still alive—the segments appear pressed together so that the sides in contact with other cells are flattened, but that after a time, when presumably the cells have died, this flattening against one another disappears, each segment at any rate in the 2- and 4-celled stages becomes spherical instead of oval and touches its neighbouring segments at points only.

Fig. 38 shows the 2-segment stage of a rabbit. Each segment is flattened against the other, but this is not due to insufficiency of space within the zona radiata because the cells do not touch the zona on all sides—and I do not see how it can be explained by reference to capillarity or surface tension or cements, because such causes ought to have the same effect after death as before. It may be objected that the change from the oval closely pressed together condition is due to a lower temperature bringing about a solidification or change in consistency of the fluids of the cells of a hot blooded animal. I am not clear how far this is due to loss of heat or how far it is due to loss of vitality. By use of a hot stage I have seen the spherical segments regain their oval and mutually compressed condition. But I have not as yet succeeded in ob-

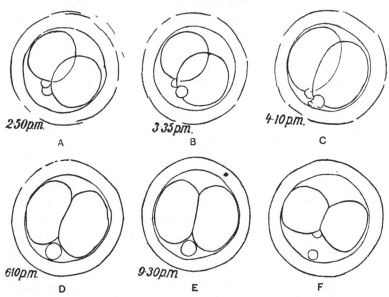

Fig. 38. *A*, ovum of rabbit 2 segment stage after removal from the Fallopian tube and examination in aqueous humour (rabbit) at the ordinary room temperature. Drawn with camera lucida. *B*, the same under the same conditions 45 minutes later. The segments are more spherical. *C*, the same after another 35 minutes during which the heat of the stage had been raised approximately to blood heat. The segments had regained their oval condition and more closely pressed together though internally there was a distinction between an inner granular and outer clearer part not noticeable in the egg when first seen in the Fallopian tube. The polar bodies showed "amoeboid movements." *D*, the same two hours later during which time the high temperature had been maintained. The outer clearer ectoplasm has diminished in extent. *E*, the same three hours twenty minutes later; the general appearance is now quite normal. *F*, the same some twelve hours later during which time the temperature had sunk to that of the room.

serving the division of a segment so cannot say whether the change is due to physical condition change, or to a re-attainment of vitality.

Przibram, when writing of the significance of the processes of cell division, refers to the analogy between the mitotic and certain other figures thus. "The mitotic forms have often been compared (Fol, Ziegler, Gallardo,) to lines of force in an electric or magnetic field, but this has not furnished any deeper analogy; in contrast to the electric and magnetic figures with dissimilar poles we are dealing in the case of mitosis with similar poles; this may be recognised from the migration of the chromosomes in opposite directions, and still more clearly from the association of several poles to form figures which do not in any way disturb the spindle and aster figures. The crossing of the rays has been urged as an objection to this comparison."

Przibram clearly favours a solution of the problem along the line of differential hydrostatic states, the accumulation and withdrawal of water from various parts of the alveolar protoplasm and he suggests as "a provisional solution of the difficult problem of mitotic cell division" the following statement: "The common cause of the mitotic phenomena lies in a localised secretion ('condensation') of a more liquid substance, the Enchylemma, and in the transformation, caused by the redistribution of the liquid, of a monocentric system of surface tension into a dicentric system."

Przibram does not allude to the paper by Marcus Hartog in the *Proceedings of the Royal Society*, 1905, Vol. LXXVI, where that author meets at any rate some of the objections alluded to above, such as the anastomosing and crossing of the rays, which he finds actually were reproduced in his magnetic model.

Hartog also obtained, though not very perfectly, triaster and tetraster conditions, although working with unlike poles of the magnet and a "third of zero sign—in the models the core of a magnet without a coil."

Still, these experiments, although they tend to make the analogy between this force of the dividing cells [termed "karyo-kinetic force," or "mito-kinetism" (Hartog)] closer, leave the analogy an analogy only, indicating that the effect upon cell material is in its action from a centre similar to that of a force like magnetism or electricity on substances variably permeable to those forces acting in a field from a centre.

Whatever the nature of the force may be, if it act under those conditions from a centre, analogous effects may be produced quite apart from any exact or near similarity existing between the supposed forces.

The explanation of invagination during gastrulation in *Amphioxus* is put forward as a little further evidence in favour of the existence of some force which acts like gravitation or magnetism or statical electricity, and can be described as acting from a centre, but which is probably of a different nature and with different laws, and is an attribute of living matter alone.

For it has been made clear, I hope, that *Amphioxus* in its process of gastrulation fulfils all the conditions required by the model of indiarubber balls, provided it can be conceded that (1) there is an attractive force between cell and cell, "acting from a centre"; (2) that the nucleus approximately indicates the position of the centre from which this force may be described as acting; (3) invagination can only take place after a certain size relative to the size of the unsegmented ovum, the distribution of yolk, and degree of force of attraction has been attained by the blastula.

By reason of the original structure of the ovum the centres of attraction take up their required position whereby invagination must ultimately occur. So long as the essential structure is not interfered with, the ovum may be subdivided without destroying the power to invaginate.

A gastrula $\frac{1}{16}$ the normal size may be obtained so long as the subdivision of the ovum has been always by a vertical plane. But the gastrula stage is not reached if the division has been made horizontally.

This fact receives satisfactory explanation on the hypothesis enunciated above.

The attraction of daughter cells immediately following separation is of course very marvellous; because whatever may be the details and however complicated may be the mechanism of cell division, yet the process of cell division reduced to its simplest expression must imply the origin of bipolarity or a bicentred condition involving an antagonism or repulsion between two centres within a body which at one time had been single centred. That is to say, the incipient daughter nuclei (and centrosome of course) while still forming parts of their disappearing parent seem to repel one another, and yet immediately on separation (which separation is by no means always nor even perhaps ever immediately absolute) appear to attract one another strongly.

But such a state of things is not unknown in other analogous phenomena in which attraction and repulsion form the visible effects of the internal force. In a bar magnet there are strong force "centres" of the two poles, while midway between the poles there is but little attractive force. Break the magnet and immediately strong attraction centres occur near the point where before there was none or but little, even if the broken ends are in loose contact. I do not mean to compare the supposed form of energy in any way with magnetism except by analogy; for magnetism is always bipolar while the form of energy under consideration appears to be automatically unipolar and bipolar alternately.

PART II

If the account given here of the gastrulation of *Amphioxus* is correct, it will be realised that the formation of the gut cavity and the definitive ectoderm and endoderm at any rate of the anterior part of the future animal is the direct result of and indeed part of segmentation. The product is on the whole a radially symmetrical animal. It is what I called in earlier papers on the Rabbit the result of the primary centre of growth (and later by paraphrase protogenesis). It is to be sharply distinguished from growth in length, which by the origin of a secondary growth-centre system (deuterogenesis) round the lips of the blastopore, converts the radially symmetrical organism into a bilaterally symmetrical one by adding on a metamerically segmented trunk and tail. In a similar way I attempted to show that the archenteron and definitive ectoderm and endoderm are formed as a direct result of protogenesis in the Amphibian, v. "The Growth in Length of the Frog Embryo."

The whole process of formation of the gastrula with its archenteron is regarded as being the simple working of the driving power, so to speak, of this supposed form of energy guided by the extrinsic factors, in this case chiefly food yolk granules which are distributed in larger or smaller grains in greater or lesser quantities within the meshes of the living protoplasm in the different regions of the egg.

In the paper alluded to before, I have attempted to show how the process may work in an Amphibian egg, but a mathematical demonstration alone would prove or disprove its correctness.

In the developing egg of *Rana temporaria* we see a gradual conversion of large segments into small segments, there being always a rather sudden transition from the smaller, usually darker cells, into the larger, usually less pigmented cells; this not very well defined transition area forms what we might call a differentiation zone gradually passing from the upper towards the lower pole. There is no migration of cells, it is merely a wave like progression of a zone of differentiation. After passing the equator one arc of the zone instead of progressing over the surface dips inwards, giving rise to a deep groove and then a cleft—the first sign of the blastopore.

A section taken vertically through the egg and this groove shows the same rather sudden mergence of the smaller dark cells into the larger light cells at the bottom of this groove just as it appears on the surface at the opposite side of the zone.

The part of the zone still on the surface sweeps on but adjoining the spot where the groove is, more and more of it turns inwards until at last the whole zone of mergence disappears from the surface but may be found beneath, where it is probably still extending and giving rise by differentiation to gut lining cells and causing a cavity by a split, which widens later into the true archenteron. This departure inwards—which does not involve any movement of cells—is shown in the accompanying figure (Fig. 39), and constitutes protogenesis, or as I called it in the paper alluded to, the result of the primary centre of cell activity. The process is somewhat masked by another phenomenon, the growth backwards of the rim of the blastopore *Bl* as soon as and wherever formed, which ultimately produces the growth in length of the animal, and metamerically segmented region, i.e. deuterogenesis (v. arrow in drawing). The ventral part dies out as an active proliferation area very soon, and the dorsal part continues and gives rise to the tail. Could not the mathematician tell us exactly, given

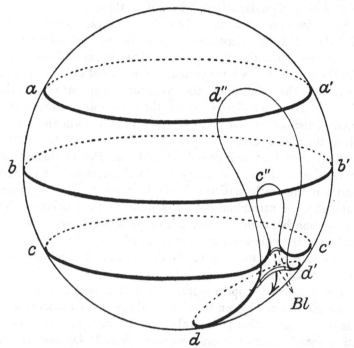

Fig. 39. An attempt to explain graphically the origin of the true archenteron in the Anuran egg by cleavage as the direct action of the same forces which produce the true archenteron in *Amphioxus* by invagination. *a–a'* a line which represents a zone of mergence of small and large segments (the effect of cell division in the horizontal plane) at an early stage of segmentation [cf. Morgan and Umé Tsuda, *Q. J. M. S.* Vol. xxxv, Pl. 24, Fig. 111]. *b–b'* the zone at a later stage. *c–c'–c"* the zone at a still later stage after gastrulation has begun. The zone has progressed superficially until after passing the equator; then at one point it has left the surface and passed inwards and backwards *c"*. *d–d'–d"* the zone still later. It is now about to leave the surface at all points and to complete the ventral lip of the blastopore. The split caused by its ingression below the surface becomes archenteron—as soon as the sides of the cleft separate. The "lips" formed by the ingression, become *ipso facto* the secondary growth centre, because, whereas before the turning in, any increase in bulk of the superficial layer (i.e. the blastula wall) leads to mere increase in the size of the radially symmetrical form, after the ingression any increase in size must on account of the folding or doubling of the superficial layer on itself bring about a conversion of the radial to the cylindrical form and thus initiate growth in length (deuterogenesis). This must be true of all gastrulae.

a force such as postulated, what the distribution of the inert or impermeable material must be to produce this form? The problem is necessarily further complicated by the interaction of the numerous force centres, though probably this effect is less where yolk is abundant than where it is absent.

Some deny the occurrence of any form of invagination in *Rana*. With them I would agree, except in so far as the withdrawal of the yolk plug inwards from the blastopore (anus of Rusconi) to form the gut floor at a later period may be called an invagination. The cases of the Urodela and the Cyclostomata appear to be rather different. Here it looks as though there were at a rather earlier period a distinct inrolling of the lower pole to form the ventral wall of the archenteron—really very much like the withdrawal of the yolk plug just mentioned. Could not the mathematician decide whether this can be or must of mechanical necessity be so? Could not the mathematician decide whether, on the hypothesis of universal cell attraction put forward above, the combined effect of attraction between cell and cell could account for the alleged inrolling of the yolk mass in certain cases?

To return for a moment to the differentiation zone alluded to in the last paragraph but one.

After this has disappeared from the surface, that is to say, after the complete formation on the surface of the circular blastopore or anus of Rusconi, what is the subsequent history of this differentiation zone? Does the differentiation cease? Or does it continue till it dies out by coalescence of all parts of the zone, during which time a complete layer is differentiated as gut epithelium? That it continues for a while, causing a split as suggested in my paper, 1894, I am confident, but it is a moot point whether it is not in some cases brought to an end by the confluence of the split produced thereby with the original segmentation cavity as maintained by Hertwig, etc.

Brachet's careful work on *Siredon* and *Rana temporaria* makes it probable that confluence is the rule in the former as it seems to be in the Gymnophiona (Brauer) and may or may not occur in the latter. But in either case the true archenteron is of protogenetic origin and is the immediate result of the simple working of forces above postulated, guided by extrinsic factors. Probably this could be demonstrated mathematically.

So also in *Amphioxus* the gut cavity is brought into being by no other means than the working of the same universal force but guided by differently arranged extrinsic factors (chiefly yolk), and by the interaction of the differently arranged intrinsic factors—the individual systems of the supposed form of energy.

There is no mysterious determinant, id or engram which in the one case causes the gut cavity to be formed by splitting, in the other by invagination, in a third case, e.g., mammal, by infiltration of fluid combined with splitting.

The relationship of the special area of cell activity, which in the vertebrate embryo gives rise to the growth in length (deuterogenesis), to the primary area (protogenesis) again most probably is capable of almost exact elucidation by mathematical means. In my papers on the Rabbit there will be found figures in which this relationship is roughly indicated, but they lack conviction, because, if for no other reason, they lack mathematical proof.

There is certainly no mammal, probably no animal, better suited for the study of this particular point than the Rabbit, because the Rabbit embryo develops from the first to the eighth day entirely free from the influence of the uterus, and is almost devoid of yolk, so that we are there dealing almost exclusively with the living matter and its properties. The zona radiata is in this case an important extrinsic factor, but this is wholly outside the living embryo.

If we get the relationship of the deuterogenetic to the protogenetic part of the organism clearly and mathematically defined, should we have an explanation of the origin of the former? Or should we have to fall back upon some supposed determinant, id or engram?

Research might give it by showing how it is in reality present from the first, though not apparent in the earlier stages (e.g., in Rabbit 1–5 days though obvious during the 7th day). If so, then there is no need for a special id or other determinant which induces at the right moment the growth in length.

A little reflection makes one realise that, with reference to those animals in which an archenteron is produced by invagination or by a process such as that characteristic of the Anura whereby a blastopore and "blastopore lips" are formed, deuterogenesis is initiated by the very fact of gastrulation, so that in those cases one can speak of deuterogenesis being the direct and inevitable consequence of the general structure of the unsegmented egg. It depends upon the doubling of the expanding area upon itself.

This may be recognised more clearly by a glance at the diagram, Fig. 40.

Let *A* represent a section of a sphere composed of living cells,

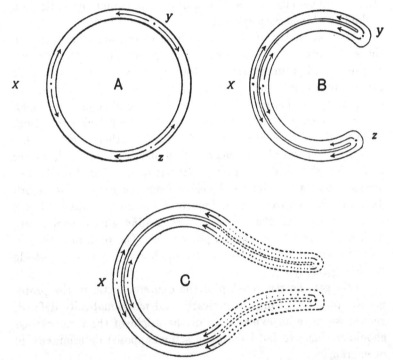

Fig. 40. *A*, a diagram representing the blastula stage. *B*, the gastrula. *C*, an embryo after growth in length has begun, the effect of which is indicated by the dotted lines. Theoretically closure of the blastopores should precede accumulation. Actually the two processes go on together as in the diagram.

which are actively dividing. The divisions occur only, or at any rate chiefly, in the radial plane and, we will imagine, with equal rapidity at all points. If this occurs in conjunction with the growth of the divided cells to their "normal" size, the result must be general increase in the size of the ring, or sphere, of which the ring is supposed to be a section. The radial symmetry is in no wise necessarily disturbed. For, given any point therein, the net result of growth must be expansion in all directions from that point as indicated by the arrows. For simplicity I represent three such areas of expansion by the three pairs of arrows. Now let it be supposed that the original arrangement of the forces involved be such that when a certain size has been attained (as in the description of the invaginating gastrula of *Amphioxus* discussed in the earlier part of this essay) invagination *must* occur, then as a result, the proliferating areas are doubled upon themselves, Fig. 40 *B*, and at the point of doubling the resultant effect must be a radial accumulation of cells and the consequent destruction of the radial symmetry, and the origin of growth in length (accompanied by at any rate partial closure of the blastopore), which will be continued so long as the margins of folding retain their power of production of undifferentiated cells. This is shown to be so in the diagram by the direction of growth which before invagination has a counterbalancing effect and so the *status quo* or radial symmetry is retained, but, as soon as an antagonism is converted into a co-operation and the balance is upset, the blastopore must tend to close and on the accumulation of radially arranged material growth in length ensues. That is to say, deuterogenesis is as inevitable a consequence of the doubling of the proliferating area as the doubling is, on my hypothesis, an inevitable result of the original general structure of the egg and it requires no special determinant to call it into being.

I confess that with regard to the origin of the deuterogenetic centre in Amniota it is not so simple a matter. In Amniota, as I believe, there is no true blastopore. In a Rabbit for instance the deuterogenetic centre seems to arise "of its own accord." For the moment the problem must remain unsolved, but further observation and reflection may show it to be as inevitably a consequence of what has gone before, as it is in the case of gastrulas with typical blastopore.

Clearly this explanation of the origin of the deuterogenetic centre, is, if correct, applicable to all gastrulae formed by invagination, or indeed, all that have a distinct blastoporic lip. If an organism retains a radial symmetry after gastrulation, it must do so by the dying out of the deuterogenetic centre of cell proliferation. By a retention of one part, and suppression of another different types of bilaterally symmetrical forms will be attained. It is characteristic of the Chordata (excepting probably the Enteropneusts) that the ventral part of the deuterogenetic centre dies out early, so that growth in length in that phylum is chiefly attributable to the dorsal part of the original blastopore lip. This dorsal part is the part of the lip, which is earliest formed in Amphibians, and as it begins its growth at once, before the whole archenteron is completed, the correct understanding of the process has been a matter of some difficulty, and has been arrived at only after much research and not a little controversy.

No doubt the denial of ids and engrams during development only puts back the difficulty to the previous generation, to the building up of the germ cells. But there is this difference. The potter turns out jar after jar exactly alike not because the clay contains an id or engram that causes the special turn of the lip or width of neck to appear at the right time, but because the mould is the same in each case.

The development of the individual is the building up of the mould which forms the next generation (i.e., the germ cells), which in most cases is all that is concerned, but in mammals at any rate the moulding is continued to a much later period of the succeeding generation's span of existence.

The attraction and repulsion observed between cell and cell are certain of the manifestations of this supposed form of energy —but probably not by any means all; just as attraction and repulsion are manifestations of electrical energy under certain conditions, but are not by any means the only manifestations. In nerve impulses we may, for instance, really be experiencing manifestations in another way of the same form of energy which under other conditions produces the attractions and repulsions and the figures of strain in the dividing cells, and the actual cell division.

Herbst made the remarkable observation that the segments of developing *Echinus* eggs and others when allowed to develop in sea water which has been deprived of calcium salts, separate from one another, instead of adhering; and that although they live so long as to develop cilia as in a fully formed blastula, nevertheless they do not form a completely closed vesicle. This observation perhaps tends to show how important this attraction property is for the normal development of embryos. The fact that under these peculiar conditions the living blastomere does not show attractive properties does not necessarily upset the hypothesis that attraction is a general truth, any more than the fact that although a glass vesicle sinks in ordinary water, yet, if a certain percentage of some salt be added to the water it exhibits repulsion rather than attraction to the centre of the earth, upsets the universality of the force of gravitation. In the latter case a better knowledge of the laws of gravitation enables one to understand the "abnormality," whereas the abnormal behaviour of segments developing in sea water without calcium salt remains for the present a mystery. It has been suggested in this particular case that the action of the calcium salt is to help in the formation of a cementing material which in its absence is not formed. This however seems to me hardly a sufficient explanation, for the presence of a cement would not of itself account for the flattening of the segments against one another, nor for the fact that on adding again the necessary calcium, adherence follows.

The idea of Entelechy developed by Driesch in his Gifford Lecture is of the most intense interest; but it must be allowed that the conception is almost mystical.

Now, although the development of the egg up to the formation of the archenteron is, as compared with the later stages of ontogeny or still more as compared with regenerative processes, almost infinitely simple, yet nevertheless it involves stages of the greatest importance, namely establishment of the gut cavity, and differentiation of the tissue into the groundwork of the ectodermal, and endodermal tissues, that is to say, the formation of layers whose subsequent fate can be foretold.

If Entelechy is the ruling influence of life, ought it not to be the ruling influence in such a process as the gastrulation of the *Amphioxus* by invagination of a blastula? If it can be shown that this process is explicable by a general application of a simple force in combination with other well-known factors, the probability of which is claimed to have been shown in the foregoing pages and elsewhere not only in *Amphioxus*, but in *Lepus* and *Rana*, may we not doubt whether, if it were possible to analyse the almost infinitely more complex processes of later development, we should find so mystical an explanation as entelechy necessary? In other words, may not this simple force be the sole factor in the development of organisms which is to be regarded as "vitalistic," that is to say peculiar to living matter—exhibiting the essential character which distinguishes living from non-living substances.

For by this supposed form of energy, I do not mean a mysterious metaphysical influence, but a form of energy comparable to gravity, electricity, or magnetism—in some respects similar to these but in other respects differing from each, and a form which could be investigated by the ordinary methods of mensuration and computation available to the mathematician.

Professor Bateson, who was so good as to read the foregoing pages, pointed out to me (among other kind and valuable criticisms, for which I desire to express my thanks) with reference to the hypothesis of an intercellular attraction as a possible cause of invagination, that, given the fact that the pull between cell and cell is eccentric at one part of a ring, or over one area of a hollow sphere, invagination would occur if the mutual pull were due to a contraction of protoplasm just as much as it would on the hypothesis of intercellular attraction.

No doubt that is so; and it must be admitted that whereas contractility is recognised as an attribute of protoplasm, the attraction between cell and cell as indicative of a special vitalistic property is "not proven."

If contraction is the cause of invagination it is necessary to assume either:

(*a*)   that there is an early differentiation into contractile tissue for the special purpose of causing gastrulation by invagination, or

(*b*)   that there is a relationship between more contractile and less contractile protoplasm so as to exert a greater pull eccentrically over a certain area as regards the cellular units corresponding in effect to the result of the forces postulated by the attraction hypothesis.

It must be admitted, that there is nothing visibly present which suggests a differentiation into more and less contractile protoplasm.

The contraction hypothesis necessitates a cement, or other firm means of attachment between cell and cell, or the units would be pulled asunder.   So far as I know invagination to form a gastrula is never spasmodic.   There are never any perceptible contractile movements.   On the other hand, the attraction hypothesis requires no special teleological developments but presupposes a constant and universal principle which, guided by other factors, in this case largely inanimate, leads inevitably under normal conditions to a certain definite result.

With a view to determining the nature of the alleged attractions between cell and cell such as have been stated to occur between the isolated blastomeres of gastrulating frogs' eggs, I repeated some of Roux's experiments by isolating the blastomeres of *Rana temporaria* and *Bufo vulgaris* and observing any changes in position or shape which take place when such cells are floating in ordinary water, or in a mixture of salt solution and albumen as used by Roux.   On the whole I found the attraction to be less noticeable than I had expected.

In many cases, that is to say between many pairs of cells which were closely situated to each other, there was no movement at all. This Roux also found.   But on the other hand certain cells originally a short distance apart from one another—perhaps half a cell diameter—certainly approached one another and came actually in contact; but I did not see any that became pressed together so as to be flattened, indicating any considerable strength of attraction force.

I experimented on frogs' eggs which were laid on the 27th March, and on toads' eggs a fortnight or so later in April.

They had been some time in dishes on my balcony, and were in the gastrulating stage, the lower lip of the blastopore was just about to form.   I broke the eggs and gently teased the cells apart, some in the water in which they were and some in a mixture of $\frac{1}{2}$ per cent. salt solution and filtered white of egg after Roux's instructions.

With the former I was much more successful in observing attraction movements, with the latter I was more successful in seeing pseudopodial movements.

After treatment of an egg as described a large number of the cells are isolated, though many pieces of gastrula wall remain unteased.

Many also are broken; but floating among the *débris* are hundreds of isolated cells as well as pairs and aggregates of three and more. Many of the cells are quite spherical as soon as one gets the preparation under observation. Other cells are oblong, or compressed in some way or other, and are seen to assume slowly a more and more spherical condition. Some of the pairs are so placed that the individuals touch one another at a point only, others are flattened against one another. The latter are probably cells in the act of division.

I have seen isolated cells which are quite close to one another come in contact, but I have not seen them flatten up against each other.

Isolated cells which are farther apart sometimes show less rapid and less regular attraction movements. For instance: there were two cells separated from each other by 6·5 divisions of the micrometer. The one on the left, the larger of the two, moved only slightly; the one on the right, the smaller, moved considerably, so that eventually they were only 2·0 divisions apart. The movements were of two kinds, slow sliding movement and sharp jerks.

But many of such movements, obviously, need have nothing to do with attraction and are amoeboid.

There are undoubted changes of shape in the isolated cells. The first tendency on separation is towards the assumption of a spherical form. This is followed by more or less marked changes in shape, sometimes so great as to amount to the protrusion of pseudopodia. So that it is probable that the slower and less certain movements are due to amoeboid activity, and the approximation of two cells may be fortuitous. And if the protrusion of pseudopodia is evidence of the contractility of some part of the cytoplasm of the cell, then clearly the cells must at this time be capable of contractile movements, which if applied in a particular way would bring about the invagination of the blastula as Professor Bateson suggests.

Thus in Fig. 41, two isolated blastomeres from a toad's egg in the gastrulating stage are shown as they appeared at intervals between 5.30 p.m. and 7.30 p.m. on April 15. In this case the slight reduction in the distance between the two cells must almost certainly have been due to the amoeboid movements exhibited by each, and in respect of one, these movements were so great as to result in well marked pseudopodia.

In Fig. 42, which is a series of drawings of three isolated cells from gastrulating toads' eggs similar amoeboid changes are seen and a slight approximation between two of the cells, *a* and *b*, has been effected.

Now these changes in shape of the isolated cell indicate in all probability contractility, at least that is one explanation of amoeboid movement, and so show that the power to exert a pull by contraction is not out of the question.

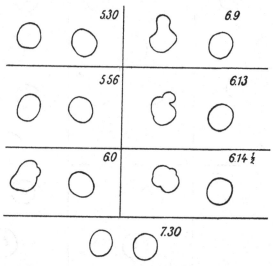

Fig. 41. Seven drawings of two isolated cells of a Toad's egg in the gastrulating stage, drawn with a camera lucida at intervals as indicated.

On the other hand there are movements, seen especially when the preparation is first made, which cannot be ascribed to any form of amoeboid activity, and as far as I can see not open to explanation by contractility.

Again, it may be that the attraction between cell and cell is greater immediately after division than at a later period. One reason why such movements occur more obviously immediately after the preparation is made than later is probably to be sought in the tendency of the isolated cells to adhere to the slide, which friction may be sufficient to prevent movement due to intercellular attraction.

If there is an attraction between cell and cell as Roux and others have suggested, that intercellular attraction is a factor which must be taken into account in the study of the mechanics of development. It is probably only in the more delicate tissues of the early stages that the attraction force can have any mechanical effect. In later stages and in the more bulky tissues its effect

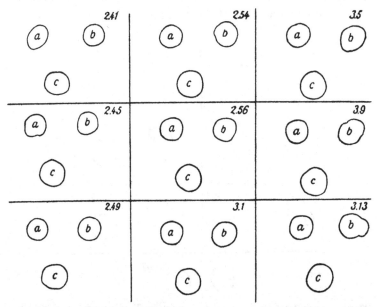

Fig. 42. Nine drawings of three cells isolated from a Toad's egg in the gastrulating stage drawn under a camera lucida at intervals of about four minutes as indicated by the figures.

would be masked by many other factors. It is quite conceivable that *contraction*, at least in its primitive condition as an attribute of protoplasm and *attraction* may be phenomena due to the same cause as, for instance, contraction of an envelope like the earth's atmosphere is due to the same cause as attraction of the earth's satellite, namely to gravitation.

Whether invagination of the *Amphioxus* gastrula and of other gastrulas of a similar type is due to the existence of an intercellular attraction or to an intracellular contraction (if not to some other cause altogether) the geometrical relationship of the nuclei considered in either connection is a matter of some interest.

While strongly inclined to adhere to the attraction hypothesis because it is simpler and more fundamental in its application I fully admit the force of Professor Bateson's criticism.

I admit quite frankly that I am advocating a vitalistic explanation for the prime biological phenomenon, namely cell division. By this I mean a form of energy evolved by and peculiar to the complex nature of the molecule of protoplasm—or of protoplasms—which exhibits so long as it retains the power of persistence an unceasing recurrence of a bipolar state out of a unipolar state which since the material is fluid induces the separation of the material aggregated round each pole, thereby multiplying the units and tending towards the establishment of more and more complex systems.

At so early a stage of inquiry it is impossible to determine how the bipolar state is to be derived from the unipolar state. Possibly it may be always bipolar, but when the energy is at its lowest ebb, namely when the cell is, as we say, in its resting stage, the poles are so close to one another as to be indistinguishable and to act in respect of external bodies as a single pole. But when the evolution of the energy grows in intensity as the result of metabolic activity the poles are driven further and further apart owing to the fluid nature of protoplasm until their separation occurs, whereupon each mass becomes instantly bipolar; but now the storm is over and the two poles in each daughter cell or nucleus lie so close to one another as to be indistinguishable. It is interesting to remember that in some cases the centrosome (which may when present mark the position of the pole or poles) appears to be doubled almost as soon as the daughter nucleus is formed.

Perhaps, however, this may be altogether too fanciful; but my point is, if we are to have a vitalistic theory of biology I submit that it must be sought for somewhat along the lines here indicated rather than in the more mystical form presented by Driesch's conception of Entelechy, though Entelechy might be a coordinating complex of this and other forces.

It is with the hope of gaining the attention of those who are mathematicians as well as biologists that I venture to offer this essay for publication.

My sincere thanks are herewith tendered to Professor Bateson, Mr Blackman, and Dr Jenkinson who were so good as to read the draft of the above essay, for their trouble in doing so, and for their criticisms.

LITERATURE REFERRED TO

ASSHETON, R. (1894). "The Primitive Streak of the Rabbit, the causes which may determine its Shape, and the part of the Embryo formed by its activity."
Quart. Journ. Micr. Sci. Vol. XXXVII.

—— (1894). "On the Growth in Length of the Frog Embryo."
Quart. Journ. Micr. Sci. Vol. XXXVII.

—— (1894). "A Re-investigation into the early stages of the Development of the Rabbit."
Quart. Journ. Micr. Sci. Vol. XXXVII.

—— (1896). "An Experimental Examination into the growth of the Blastoderm of the Chick."
Proc. Roy. Soc. Vol. LX.

—— (1898). "The Development of the Pig during the First Ten Days." *Quart. Journ. Micr. Sci.* Vol. XLI.

—— (1898). "The Segmentation of the Ovum of the Sheep, with observations on the Hypothesis of a Hypoblastic Origin for the Trophoblast." *Quart. Journ. Micr. Sci.* Vol. XLI.

—— (1905). "On Growth Centres in Vertebrate Embryos."
Anat. Anz. Vol. XXVII.

—— (1907). "Certain Features Characteristic of Teleostean Development." *Guy's Hospital Reports.* Vol. LXI.

—— (1912). "Gastrulation in Birds."
Quart. Journ. Micr. Sci. Vol. LVIII.

BALFOUR, F. M. (1885). "Vertebrate Embryology." Macmillan.

BATESON, W. (1884). "The Development of Balanoglossus."
                     Quart. Journ. Micr. Sci. Vol. XXIV.

VAN BENEDEN, E. (1912). "Recherches sur l'Embryologie des
    Mammifères."   *Arch. de Biol.* Vol. XXVII.

BRACHET, A. (1903). "Recherches sur l'ontogenèse des Amphibiens,
    Urodèles et Anoures (Siredon pisciformis—Rana temporaria)."
                                  Arch. de Biol. Vol. XIX.

BRAUER, A. (1897). "Beiträge zur Entwickelungsgeschichte der
    Gymnophiona."   *Zool. Jahrbuch.*

CERFONTAINE, P. (1907). "Recherches sur le développement de
    l'Amphioxus."   *Arch. de Biol.* Vol. XXII.

CONKLIN, E. G. (1905). "The Orientation and Cell Lineage of the
    Ascidian Egg (Cynthia partita)."
                Journ. Acad. Sci. Philadelphia. Ser. 2. Vol. XIII.

CONN, H. W. (1884). "Development of Serpula."
                                    Zool. Anz. Vol. VII.

DOHRN, A. (1875). "Der Ursprung der Wirbeltiere und das Princip
    der Functionswechsels." Leipzig.

DRIESCH, H. (1908). "The Science and Philosophy of the
    organism." (The Gifford Lectures. 1907, 1908.) London.

DUVAL, M. (1889). "Atlas d'Embryologie." Paris.

GASKELL, W. H. (1908). "The Origin of Vertebrates." London.

GOETTE, A. (1884). "Abhandlungen zur Entwicklungsgeschichte
    der Thiere." Heft 2. Hamburg u. Leipzig.

—— (1886). "Untersuchungen zur Entwicklungsgeschichte von
    Spongilla fluviatilis." Loc. cit. Heft 3.

GOODALE, H. D. (1911). "The Early Development of Spelerpes
    bilineatus."   *Amer. Journ. Anat. Philadelphia.*

HARTOG, M. (1905). "The dual force of the dividing cell. Pt 1.
    The Achromatic Spiral Figure illustrated by Magnetic Chains
    of Force."
        Proceedings of the Royal Society. Series B. Vol. LXXVI.

HATSCHEK, B. (1881). "Studien über die Entwicklung des Amphioxus." *Arch. Zool. Inst. Wien.* Bd. IV.

—— (1909). "Studien zur Segmenttheorie des Wirbeltiere." *Morph. Jahr.*

HERBST, C. (1900). "Ueber das Auseinandergehen von Furchungs- und Gewebezellen in kalkfreiem Medium." *Archiv f. Entw.-Mech.* Bd. IX.

HERTWIG, O. (1906). "Handbuch der Vergleichenden und Experimentellen Entwickelungslehre der Wirbeltiere." Vol. I. Jena.

HIS, W. (1874). "Unsere Körperform und das Physiologische Problem ihrer Entstehung." Leipzig.

—— (1874). "Ueber die Bildung des Lachs-Embryo." Leipzig.

HUBRECHT, A. A. W. (1905). "Gastrulation of Vertebrates." *Quart. Journ. Micr. Sci.* Vol. XLIX.

—— (1912). "Frühe Entwickelungsstadien des Igels und ihre Bedeutung für die Vorgeschichte (Phylogenese) des Amnions." *Zool. Jahr.* Bd. III.

JENKINSON, J. W. (1913). "Vertebrate Embryology." Oxford.

KASTSCHENKO, N. (1888). "Zur Entwickelungsgeschichte des Selachierembryos." *Anat. Anz.* III.

KEIBEL, F. (1901). "Die Gastrulation und die Keimblattbildung der Wirbeltiere." *Erg. d. Anat. und Entw.-Gesch.* Vol. X.

KLAATSCH, H. (1897). "Bemerkungen über die Gastrula des Amphioxus." *Morph. Jahrb.* Bd. XXV.

KOPSCH, F. (1896). "Experimentellen Untersuchungen über den Keimhautrand der Salmoniden." *Verh. u. Anat.-Gesel.* Berlin.

KOWALEVSKI, K. (1867). "Entwicklungsgeschichte des Amphioxus lanceolatus." *Mém. Acad. Imp. de St Pétersbourg.* T. XI.

DE LANGE, D. (1912). "Mitteilungen zur Entwickelungsgeschichte des Riesensalamanders (Megalobatrachus maximus Schlegel)." *Anat. Anz.* Vol. XLII.

LEGROS, R. (1907). "Sur quelques cas d'asyntaxie blastoporiale chez l'Amphioxus." *Mitt. Zool. Stat. Neapel.* Bd. 18.

LILLIE, F. (1908). "Development of the Chick." New York.

LWOFF, B. (1894). "Die Bildung der primären Keimblätter und die Entstehung der Chorda und des Mesoderms bei den Wirbeltieren." *Bull. de la Soc. Imp. des Nat. de Moscou.*

MACBRIDE, E. W. (1898). "The early Development of Amphioxus." *Quart. Journ. Micr. Sci.* Vol. XL.

—— (1910). "The Formation of the Layers in Amphioxus and its Bearing on the Interpretation of the Early Ontogenetic Problems in other Vertebrates." *Quart. Journ. Micr. Sci.* Vol. LIV.

MARSHALL, A. M. (1893). "Vertebrate Embryology." London.

MORGAN, T. H. (1896). "The number of Cells in Larvae from Isolated Blastomeres of Amphioxus." *Arch. f. Entw.-Mech.* Bd. III.

—— (1907). "Experimental Embryology." Macmillan.

—— and HAZEN, A. P. (1900). "The Gastrulation of Amphioxus." *The Journal of Morphology.* Vol. XVI.

PATTEN, W. (1912). "The Evolution of Vertebrates and their Kin." Philadelphia.

PATTERSON, J. T. (1909). "Gastrulation in the Pigeon's Egg. A Morphological and Experimental study." *Journ. of Morph.* Vol. XX.

PRZIBRAM, H. (1908). "Embryogeny." University Press, Cambridge.

REINKE, F. (1900). "Zum Beweis der trajectoriellen Natur der Plasmastrahlungen." *Archiv f. Entw.-Mech.* Bd. IX.

RHUMBLER, L. (1902). "Zur Mechanik des Gastrulationsvorganges, insbesondere der Invagination."
Archiv f. Entw.-Mech. Bd. XIV.

ROUX, W. (1894). "Ueber den 'Cytotropismus' der Furchungs-zellen des Grasfrosches (Rana fusca)."
Archiv f. Entw.-Mech. Bd. I.

—— (1896). "Ueber die Selbstordnung (Cytotaxis) sich 'berüh-render' Furchungszellen der Froscheier."
Archiv f. Entw.-Mech. Bd. III.

RÜCKERT, J. (1899). "Die Erste Entwicklung des Eies der Elasmobranchier." Jena.

SAMASSA, P. (1898). "Studien über den Einflusz des Dotters auf die Gastrulation und die Bildung der primären Keimblätter der Wirbeltiere. IV. Amphioxus."
Archiv f. Entw.-Mech. Bd. VII.

SEDGWICK, A. (1884). "On the Origin of Metameric Segmentation and some other Morphological Problems."
Quart. Journ. Micr. Sci. Vol. XXIV.

—— (1886). "On the Fertilised Ovum and the Formation of the Layers of the S. African Peripatus."
Quart. Journ. Micr. Sci. Vol. XXVI.

SEMPER, C. (1874). "Ueber die Stammesverwandtschaft der Wirbeltiere und Anneliden." *Cent. f. Med. Wiss.*

SHEARER, C. (1911). "On the Development and Structure of the Trochopore of Hydroides Uncinatus (Eupomatus)."
Quart. Journ. Micr. Sci. Vol. LVI.

SOBOTTA, J. (1897). "Beobachtungen über den Gastrulations-vorgang beim Amphioxus."
Verh. d. Physikal.-Mediz. Ges. zu Würzburg. Bd. XXVI.

STOSSICH (1878). "Beiträge zur Entwickelung der Chaetopoden."
Sitz. d. k. k. Akad. Wiss. B. 77– .

THÉEL, HJALMAR (1892). "The Development of Echinocyamus pusillus."
N. Acta Reg. Soc. Sc. Upsala. Na. Ser. Abstr. in *Nature.*
Vol. XLVIII. p. 330.

TREADWELL, A. L. (1901). "The Cytogeny of Podarke obscura."
Journ. of Morph. Vol. XVII.

WILL, L. (1892). "Beiträge zur Entwickelungsgeschichte des
Reptilien."   *Zool. Jahr.* Vol. VI.

WILLEMOES-SUHM, R. V. (1871). "Biologische Beobachtungen
über niedrige Meeresthiere." *Zeit. f. Wiss. Zool.* Vol. XXI.

WILSON, G. B. (1892). "On multiple and partial Development in
Amphioxus."   *Anat. Anz.* Vol. VII.

—— (1893). "Amphioxus and the Mosaic Theory of Development."   *Journ. of Morph.* Vol. VIII.

ZUR STRASSEN, O. (1903). "Ueber die Mechanik der Epithelbil-
dung."   *Verh. d. deutsch. zool. Ges.* S. 91–112.

Printed in the United States
By Bookmasters